Bio-inspired Materials and Sensing Systems

# *Bio-inspired Materials and Sensing Systems*

**Peter Biggins**
*Dstl, Porton Down, Salisbury, Wiltshire, UK*

**John Hiltz**
*DRDC Atlantic, Halifax, Nova Scotia, Canada*

**Anne Kusterbeck**
*Center for Biomolecular Science and Engineering, US Naval Research Laboratory, Washington D.C., USA*

RSC Publishing

ISBN: 978-1-84973-121-8

A catalogue record for this book is available from the British Library



Published by The Royal Society of Chemistry,
Thomas Graham House, Science Park, Milton Road,
Cambridge CB4 0WF, UK

Registered Charity Number 207890

For further information see our web site at www.rsc.org

# *Preface*

*'Nature only knows transformation' Anon*

The challenges facing societies in the twenty-first century are considerable whether they are from a global warming, environmental pollution, fuel dependency or terrorism perspective. As a result there is a growing requirement for the academic, research and industrial communities to look for innovative ways of addressing these issues. It is against this context that this book not only explores how nature's paradigms can be applied to defence and security needs but also addresses some of the wider challenges facing society. *Bio-inspired Materials and Sensing Systems* explores how we can learn from Nature and use this knowledge to provide new approaches to solving some of these challenges. The idea for this book developed from a presentation addressing the theme 'learning from nature' presented at the First UK-US Conference on Chemical and Biological Sensors held at the Royal Society of Chemistry, London, UK, in April 2007.

The authors acknowledge that the topics covered in this book only scratch the surface of a considerable amount of research that has been completed and is currently ongoing in a wide variety of disciplines. Yet there is a clear need for interdisciplinary or cross-cutting interactions to truly benefit from the work being undertaken.

We are indebted to a large number of colleagues for discussions and inputs that led to the writing of this book. They include John Lewis, Martin McDonnell, John Luong, Lars Piehler, Steven Nicklin, Alistair Hutchinson, Andy Bell, Doug Imeson, Georgina Biggins and Mike Steeden.

Bio-inspired Materials and Sensing Systems
By Peter Biggins, John Hiltz and Anne Kusterbeck

Published by the Royal Society of Chemistry, www.rsc.org

Finally, we gratefully acknowledge our families, especially our spouses Caroline, Crystal and Bill, for all their patience, help and understanding while we completed this book.

Peter Biggins
DSTL
John Hiltz
DRDC
Anne Kusterbeck
NRL

# Contents



Bio-inspired Materials and Sensing Systems
By Peter Biggins, John Hiltz and Anne Kusterbeck

Published by the Royal Society of Chemistry, www.rsc.org

CHAPTER 1

# A View on Bio-inspiration

## 1.1 Introduction

Bio-inspiration is not a new concept; the human race throughout its history has looked to the environment for inspiration. Many of the tools and concepts we use today have their provenance from the natural world, for example flight and birds, dogs' noses and sensing, and vision and cameras. However, there is a growing body of thought worldwide that bio-inspired or bio-derived approaches can provide a wealth of innovative solutions and that the applications of biotechnology will play a predominant role in the next 'technical revolution'. In fact, biotechnology and bio-inspired approaches can help meet the many challenges of defence and security, not only in such areas as situational awareness, surveillance, network centric operations and protection/performance but also in many civilian application areas.

*Bio-inspired Materials and Sensing Systems* explores how we can learn from Nature to provide new approaches to solving some of these challenges. In particular, the concerns regarding terrorism[1] are well founded as graphically demonstrated with the events of 2001 in the United States (US) and subsequently in the United Kingdom (UK) in 2005 and further incidents in various parts of the globe.

This is not an exhaustive study as the subject is extremely wide and varied and there are many researchers already carrying out excellent work in discrete areas. Furthermore, the increasing number of publications in a variety of journals illustrates the growing awareness of an interdisciplinary approach to this fascinating field. To illustrate this, an appraisal of patents published between 1985 and 2005 showed that whilst the total number of patents increased over the period examined, those with biomimetic content

Bio-inspired Materials and Sensing Systems
By Peter Biggins, John Hiltz and Anne Kusterbeck

Published by the Royal Society of Chemistry, www.rsc.org

had increased faster as a proportion of total patent publications.[2] There are also many excellent references available on specific areas (*i.e.* biosensors,[3–4] biomaterials,[5] *etc.*). Our goal is to provide an overview of how bio-inspired/-derived approaches can be used to enhance components, systems and systems of systems for defence and security applications.[6] This book aims to provide a wider perspective of ongoing work in these areas and focus on structures, sensing, power, processing and potential applications.

## 1.2 Context/Motivation

Why are we looking at the natural world for inspiration and in particular biological organisms? Biological organisms use readily available natural materials to assemble highly functionalized structures, sensing and processing systems. They have evolved packaging that is relevant and appropriate to their environments in terms of physical structure, physiology, repair, sensing and processing attributes along with a whole series of survival strategies. In addition, through successive generations, the ability to transfer genetic coding relevant to the next generations provides an adaptive capability. The motivation, therefore, is to use biotechnology, bio-inspiration and advances in our understanding of biological processes to lead toward the development of new materials and artificial sentinel systems that can confirm the exposure to a wide range of unknown toxic materials.

To fully understand and learn from Nature, the importance of an interdisciplinary approach cannot be stressed enough. In fact, interdisciplinary research in biotechnology and the related scientific areas has increased tremendously over the past decade. This pace of growth, in conjunction with advances in micro-fabricated systems, computer hardware, bio-engineering and the availability of low-powered miniature components, has now made it feasible to design bio-inspired materials, sensors and systems with tremendous potential to meet many of today's challenges. The defence and security areas are well suited to this type of approach as the output may enhance the paradigms in the way we sense and process information from the environment.

### 1.2.1 Bio-inspiration

What do we mean by bio-inspiration, especially when it is currently the fashion to put 'bio' (the abbreviation for biology) in front of a range of technical terms? Bio-inspiration is often referred to as biomimicry or biomimetics. It is therefore important that biomimicry/biomimetics and bio-inspiration be defined:

> ***Biomimicry*** (from bios, meaning life, and mimesis, meaning to imitate) is a relatively new science that studies Nature, its models, systems, processes and elements and then imitates or takes creative inspiration from them to solve human problems.
>
> Another definition of biomimicry is: a conscious strategy by designers to observe and learn principles of design from Nature.[7]

***Biomimetics*** (as already mentioned above) is the word most frequently used in scientific and engineering literature to indicate the underlying biological paradigms present keeping each species functioning in its own unique way.[8]

***Bio-inspired*** relates to ideas inspired by mechanisms or laws operating in biological organisms (*e.g.* neural networks).[9]

Other descriptors frequently encountered include:

***Bio-derived*** materials are materials made from or originating from living organisms.

***Bio-fabrication*** is a process using cells, viruses, proteins, biomaterials and bio-active compounds as building blocks to fabricate advanced biological models, medical therapeutics and non-medical biological systems.

Bio-inspiration (although purists prefer the term 'bionics') in reality is a term that is all embracing as it covers all aspects of Nature. It allows us to follow Nature by using natural or other materials such as polymers and ceramics.

From an engineering viewpoint the term biomimetic (accredited to Otto Schmitt in the 1950s[10]) has been used for describing how ideas from Nature can be used for engineering purposes. Bionics is a term attributed to Jack Steele at the US Air Force Research Laboratory at Wright Patterson Air Force Base in Dayton, Ohio. It is used to describe the application of processes, structures and systems found in Nature to engineering and technological developments.

The report on Biomimetic Engineering for Space Applications[11] notes that Gustave Eiffel was inspired by the structure of the human femur (thigh bone) when designing the flared shape of the Eiffel Tower. The design was based on observation of the lattice work of tiny ridges of bone (called trabaculae). The arrangement of the trabaculae provided an effective way of supporting the mass of the structure. The iron curves of the Eiffel Tower were based on this principle, such that the structure of the tower could withstand the shear and strain forces due to wind flow.

Currently, we refer to bio-inspiration as a model that allows us to either copy or adapt Nature's processes. Biological systems are able to sense many stimuli, such as light, sound, heat, pressure, chemicals (pheromones for example), gravitational and electric fields. Complex processing systems and physical structures have also evolved to enhance the survivability of organisms in the environment. The mechanisms underpinning these biological sensing functions, material structures and processing systems have application in a wide range of sensor technologies and structural designs and have the potential to provide capabilities that are not possible with conventional methods. These include for example: the development of receptors that are genetically engineered or are chemically synthesized to reduce cost and increase specificity with a wider dynamic operating range; the preparation of bio-inspired materials such as 2-D and 3-D imprinted polymers, protein nucleic acid scaffolds, biomimetic structures and surfaces. With the growth in bio-informatics, significant advances are also being made in genomics, proteomics and metabolomics. The incorporation of micro- or nano-fabrication technologies allows the development of low-cost microarrays, novel structures and self-assembling systems.

### 1.2.2 Biotechnology *vs.* Bio-inspiration

A key factor in the increased focus on bio-inspiration has been the rapid advances in the related area of biotechnology. Although there are a number of definitions for biotechnology, the following is used in this book:

> *Any technique that uses living organisms, or parts of organisms, to modify products, improve plants or animals, or to develop micro-organisms for specific uses.*

Though we can exploit biotechnology for specific uses, it is through our understanding of how Nature develops processes and systems that we can learn and adapt them for our needs in a wide range of defence and civilian-related applications. For example, specific biotechnology techniques developed in molecular biology and genetics have resulted in major advances in industrial processes, agriculture and medicine. These significant achievements coupled with engineering and design will allow bio-inspired ideas to become a reality.

## 1.3 Challenges for Science and Technology

This is an opportune time for developing bio-inspired approaches to some of our most difficult security issues. Whilst there is substantial investment in bio-inspiration and biotechnology (the level of investment is discussed more fully in Chapter 2), other critical questions remain:

- How can scientists and engineers replicate what Nature does? To be able to meld together collagen and chalk, both weak materials, to produce a structure that has the properties of shell or bone would be a major materials breakthrough. To be able to form these materials with minimal energy has the potential to revolutionize industrial processes.
- How can we develop systems that will operate in extreme environments in addition to the laboratory? Although significant strides are being made in the area of chemical and biological detection and identification, the need for rapid, specific and sensitive response is paramount. In particular, biological detection in the field is technologically challenging; agent levels can be low compared to a variable and sometimes high natural background;[12] tests using laboratory-based equipment, techniques and their associated reagents, though well developed, usually require qualified personnel; tests are slow and can take several hours to produce a result, which makes them largely unsuitable for field use. The challenge is to develop technologies and techniques for a fieldable detection capability for bacteria, viruses and toxins which do not respond to background material.

- Can we learn from Nature to develop systems or methodologies that will enable us to collect sufficient material for detection and identification of the agent while being able to discriminate against interferents in the time frames required?

In summary, the real question is, can we develop rugged, reliable, reproducible and integrated systems that will effectively meet the needs of the defence, security and civilian communities?

## 1.4 The Need for a Framework

The questions above point to the need for a framework to bound the problem and exploit the vast amount of research that is being carried out in the bio-inspired area. The graphic in Figure 1.1 shows a conceptual space in which biological principles can be divided into those found in natural, hybrid and non-biological (but using biological characteristics) systems. These can be used to generate applications and articulate the requirements needed to solve the scientific and technological challenges. On the baseline (from left to right) the developing science base provides the basis for the technologies that meet the requirements and applications for the mission space. Anywhere along the baseline there is a level of development that transitions between a non-biological solution (which is using biological characteristics) to a totally biologically inspired solution.

The framework provides the basis for understanding how the bio-inspired approaches may meet requirements. Examples of the categories referred to in the framework are discussed in the following sections.

### 1.4.1 Biological Principles

The ultimate vision of 'bio' focused science and technology is to replicate biological principles found in Nature. There are numerous examples in Nature

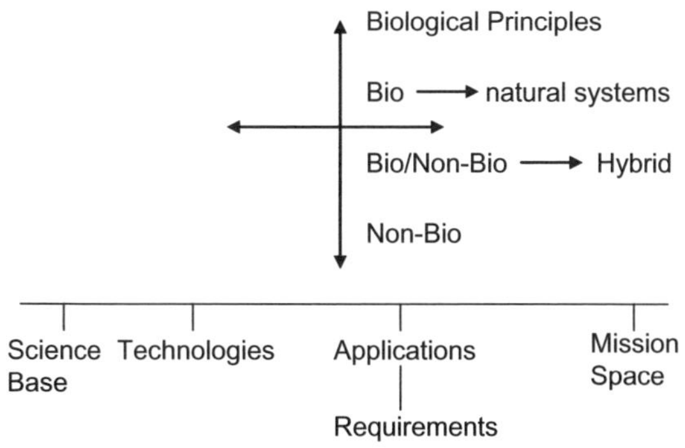
Biological Principles
Bio → natural systems
Bio/Non-Bio → Hybrid
Non-Bio
Science Base
Technologies
Applications
Requirements
Mission Space

**Figure 1.1** A conceptual framework for bio-inspired science and technology.

of the superiority of biological systems in terms of self-assembly, design, replication, use of readily available materials, energy efficiency, systems integration and collective behaviour.

#### *1.4.1.1 Natural Systems*

For detection purposes, the use of caged canaries by miners to detect methane in coal mines is an example of the exploitation of natural systems. The canary is highly sensitive to methane, an odourless and poisonous gas. The canary dying provided a visible, functional alarm to the miners that they should leave the area. Toxic substances target natural receptor sites in an organism. A function-based system (such as the canary) will react to uncharacterized substances, which, if harmful, will affect the biological function of the organism. Therefore, function-based sensing techniques could be used to provide a generic detection and classification capability for the identification of toxic substances.

Biological systems are widely used in environmental testing to provide an integrated picture of overall toxicity of an effluent or a sample of water, sediment, or soil from a contaminated site.[13] Daphnia, small freshwater crustaceans, are used to provide an indication of water quality. They are sensitive to changes in water chemistry and their response to toxicants can be readily monitored.[14] Whole cell micro-organisms, such as the unicellular green algae, *Chlorella vulgaris*, have been used to detect nerve agents. Exposure to toxicants causes changes in their characteristic fluorescence induction curves with resultant changes in photochemical yields.[15,16]

A whole organism can be monitored remotely to look for changes in behaviour that are indicators of changes in the environment. Isolated fish chromatophore cells can also be used as a detection system.[17] This approach exploits the ability of living chromatophores to respond to many active substances such as bacterial toxins; the response of the chromatophore is measured as changes in the appearance of the cell due to intracellular activity of the colorants.[18] These examples illustrate that isolated sentinel organisms have been used for many years as natural detectors/indicators and alarms.

In natural systems bees demonstrate excellent communication and knowledge transfer for collecting material, and much can be learnt about how they share information between themselves, particularly from their swarming behaviour. The work of Menzel and colleagues[19] demonstrates how honeybees can be trained to collect specific material for sensing.

#### *1.4.1.2 Bio-hybrid*

Bio-engineered cell systems that include direct binding of the analyte are under development. The advantage of this type of approach is that it makes use of the cell's built-in signal amplification system. An example of this is the MIT Lincoln Laboratory work on B cells.[20] Researchers have developed a sensor called CANARY in which they exploit B cells found in vertebrate immune systems. These cells can be cloned and engineered so they produce populations which express a single antibody that is specific for the target of interest. The process works when an antigen binds with the surface expressed antibody resulting in a release of $Ca^{2+}$ ions. The B cell also has the luminescent protein aequorin derived from jellyfish (*Aequorea Victoria*). The aequorin responds to the increased levels of $Ca^{2+}$ by emitting blue-green light.

Another example is based on a yeast cell in which the mammalian olfactory signalling pathway has been inserted. Rat olfactory proteins have been inserted into the outer membrane of yeast cells. The yeast cells can be used to detect a TNT-associated compound when the cells turn a fluorescent green.[21] The rat olfactory proteins stimulate an increased cellular level of cyclic adenosine monophosphate (cAMP). The yeast was genetically engineered to make green fluorescent protein when the cAMP switch was turned on.

Variants of this type of approach in the past have included the development of sensing systems based on ion channel proteins incorporated in synthetic membranes to increase detection sensitivities.[22] Detector systems based on genetically modified bioluminescent organisms have been successful. The coupling of genes that respond directly to the presence of various target analytes to firefly luciferase by genefusion has been demonstrated.[23]

Cells and tissues used outside their natural environment can behave quite differently from how they would if they remained in the host. Hence, whole organism measurements, if fast enough, can provide an attractive route for generic detection. However, operational deployment of cell-based sensing systems provides some unique challenges to an already overburdened logistical system. Any system that utilizes living material will have to have an accompanying life support system. The complexity of this system will be determined by the element(s) requiring support.

Another exciting area is the potential for continuous non-invasive monitoring of blood chemistry and other metabolic diagnostic indicators to follow changes due to the effects of toxicants on metabolic processes.[24]

### *1.4.1.3 Hybrid Systems*

Biosensors represent a hybrid type of system in which a bio-derived receptor is used to detect a specific target analyte, either biological/organic material or chemical vapour, and the subsequent binding effect is monitored by an engineered transducer of some type. As discussed in more detail in Chapters 6 and 7, this interaction can be modulated by substituting receptors, transducers or other hardware to allow optimal detection.

Biosensors are widely used in the commercial market, in medical, veterinary, environmental and manufacturing/processing industries. There have been significant advances in the biosensor field and development of sensors for the detection of biological warfare agents has recently been reviewed.[25,26] An example of an early biosensor used by UK forces is the Nerve Agent Immobilised Enzyme Alarm and Detector (NAIAD) (Figure 1.2) to detect nerve agents such as GA (tabun) and GB (sarin), and attack concentrations of

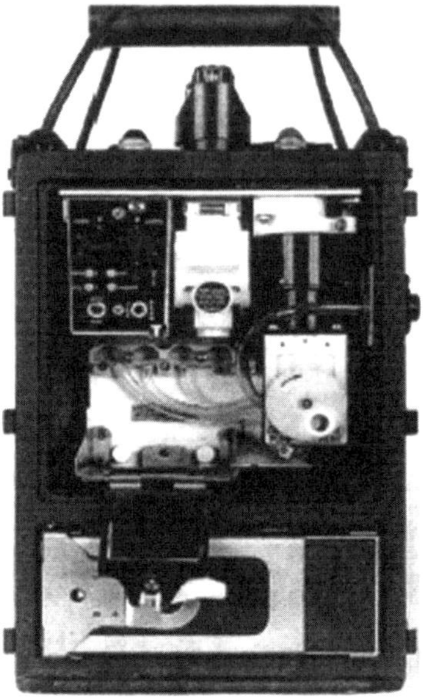

**Figure 1.2** Nerve Agent Immobilised Enzyme Alarm and Detector (NAIAD), an example of an early biosensor used by the UK forces to detect chemical agents.

hydrogen cyanide.[27] The bioreceptor in this case is the enzyme butryl cholinesterase, which shows the same activity as that of human acetyl cholinesterase. Whilst this type of detector is extremely sensitive to low levels of nerve agent and HCN, it does rely on wet reagents and regular resupply. This illustrates one of the issues associated with biosensor systems in terms of the demand for consumables and the attendant support system.

#### *1.4.1.4 Non-biological*

Non-biological systems are those that are synthetic and that incorporate bio-derived strategies, structures and processing architectures. For example, there is a growing need for sensing systems to be mobile. Bio-inspired robotic systems are being developed which not only carry a package of sensing elements but are intended to harvest energy from the natural environment. An example of such a concept robotic system is SlugBot[28] which has been designed to identify and collect slugs with the aim of delivering them to a collection point where they will be digested to produce energy with which it can recharge itself. The researchers that demonstrated SlugBot are now developing a series of 'bots' called EcoBot 1, 2 and 3, which will use on-board microbial fuel cells to provide power.

At its Biorobotics Laboratory, Carnegie Mellon University, researchers are exploiting snake motion to develop robotic systems with mobility along, through or over a variety of terrains using climbing, swimming and pipe crawling capabilities.[29] NASA is developing intelligent robot snakes that can be used to explore planets' surfaces,[30] whilst BAE systems are developing low-cost SnakeBots for the battlefield.[31] It is becoming feasible to develop complete, bio-inspired robotic systems with adhesive capabilities based on the lumpsucker fish, which has a powerful suction disk formed by the control of pelvic fins. Another example is the exploitation of the way in which a gecko climbs walls by controlling van der Waals forces between its pads and the climbing surface.[32]

## 1.5 Science and Technology to Mission Capability

Across the underlying research and technology spectrum, these approaches (biological, hybrid and non-biological) need to be developed to meet operational requirements. A fundamental issue is the transitioning of developments in research into new hardware for a range of missions. Operational requirements are directly linked to the mission space. For a given scenario, the technology required may be simple, stand-alone sensing or autonomous and the science to realize this can range from simple to complex. Feedback is needed as well between those defining requirements and capabilities and those involved in developing science and technology (S&T), *i.e.* as new requirements are brought forth, there is often the need to go back to the S&T community to find solutions to challenges such as making systems that improve capabilities or are cheaper, faster or more reliable. There is a critical need to understand the basic science in order to adapt and improve technologies in the field. Bio-inspiration may also provide new approaches to putting components together to make an integrated system.

## 1.6 Conclusion

Clearly, the defence and security communities have diverse requirements for bio-inspired technologies that may require different levels of investment, as addressed in the next chapter. The promise of this area is such that it cannot be ignored, both for the capabilities it will provide and to avoid technological surprise.

## References


1. Commission of the European Communities, Green paper on Bio-preparedness, Brussels, 11.7.2007, COM (2007), 399 final.
2. R. H. C. Bonser, *J. Bionic Eng.*, 2006, **3**(1), 39–41.
3. F. S. Ligler and C. Rowe Taitt (Eds.), *Optical Biosensors: Today and Tomorrow*, 2010, 2nd Edition, Elsevier Science, ISBN-10 0444531254.
4. R. S. Marks, C. R. Lowe, D. C. Cullen, H. W. Weetall and I. Karube (Eds.), *Handbook of Biosensors and Biochips*, 2007, Wiley-Blackwell, ISBN-10 0470019050.
5. B. Basu, D. S. Katti and A. Kumar, *Advanced Biomaterials-Fundamentals, Processing and Applications*, 2009, Wiley, ISBN-10 0470193409.
6. P. D. E. Biggins, A. Kusterbeck and J. A. Hiltz, *Analyst*, 2008, **133**, 563–570.
7. http://siliclone.tripod.com/books/history/H111.html

8. http://en.wikipedia.org/wiki/Biomimicry
9. http://en.wiktionary.org/wiki/bio-inspired
10. http://en.wikipedia.org/wiki/bionics
11. M. Ayre and N. Lan, *Biomimetic Engineering for Space Applications*, (Editor K. Fletcher), 2006, ESA Publictions Division, ESTEC, Noordmijk, The Netherlands, ISBN-10 9290929995.
12. R. M. Harrison, A. M. Jones, P. D. E. Biggins, N. Pomeroy, C. S. Cox, S. P. Kidd, J. L. Hobman, N. L. Brown and A. Beswick, *Int. J. Biometeorol.*, 2003, **49**, 167–178.
13. F. Botre, E. Podesta, B. Silvestrini and C. Botre, *Ann. Ist. Super. Sanita*, 2001, **37**, 607–613.
14. USEPA, Methods for measuring the acute toxicity of effluents and receiving waters to freshwater and marine organisms, US Environmental Protection Agency, Washington, DC, EPA/00/4-90-027, 1991.
15. C. A. Sanders, M. Rodriguez and E. Greenbaum, *Biosens. Bioelectron.*, 2001, **16**, 439–446.
16. M. Rodiguuez, C. A. Sanders and E. Greenbaum, *Biosens. Bioelectron.*, 2002, **17**, 843–849.
17. F. W. Chaplen, R. H. Upson, P. N. McFadden and W. Kolodziej, *Pigment Cell Res.*, 2002, **15**, 19–26.
18. R. R. Preston and P. N. McFadden, *Biosens. Bioelectron.*, 2001, **166**, 447–455.
19. R. Menzel and U. Muller, *Annu. Rev. Neurosci.*, 1996, **19**, 379–404.
20. T. H. Rider, M. S. Petrovick, F. E. Nargi, J. D. Harper, E. D. Schwoebel, R. H. Mathews, D. J. Blanchard, L. T. Bortolin, A. M. Young, J. Chen and M. A. Hollis, *Science*, 2003, **301**, 213–215.
21. V. Radhika, T. Proikas-Cezanne, M. Jayaraman, D. Onesime, J. H. Ha and D. N. Dhanasekaran, *Nat. Chem. Biol.*, 2007, **3**, 325–330.
22. B. A. Cornell, V. L. B. Braach-Maksvytis, L. G. King, P. D. J. Osman, B. Raguse, L. Wieczorek and R. J. Pace, *Nature*, 1997, **387**, 580–583.
23. M. Chalfie, Y. Tu, G. Euskirchen, W. W. Ward and D. C. Prasher, *Science*, 1994, **263**, 802–805.
24. F. Torella, R. Cowley, M. S. Thorniley and C. N. McCollum, *Comp. Biochem. Physiol., A: Mol. Integr. Physiol.*, 2002, **132**, 199–203.
25. M. F. Kramer, *Clin. Microbiol. Rev.*, 2005, **18**, 583.
26. J. J. Gooding, *Anal. Chim. Acta*, 2006, **559**, 137–151.
27. R. J. Powell, *Chem. Br.*, 1988, **24**, 665.
28. I. Kelly and C. Melhuish, *SlugBot: A Robot Predator, Proc. European Conference on Artificial Life, Prague*, 2001, **2159**, 519–528.
29. Research at the Robotics Institute: www.ri.cmu.edu/general/research.html
30. NASA developing 'Snakebot' to explore and build in space: http://astrobiology.arc.nasa.gov/news/expandnews.cfm?id = 367
31. Robot spy can survive battlefield damage, New Scientist, 20 August 2003: www.newscientist.com/article.ns?id = dn4075 (accessed 28/05/2010)
32. K. Autumn, M. Sitti, Y. C. A. Liang, A. M. Peattie, W. R. Hansen, S. Spenberg, T. W. Kenny, R. Fearing, J. N. Israelachvili and R. J. Full, *Proc. Natl. Acad. Sci. USA*, 2002, **99**, 12252–12256.

CHAPTER 2

# *Investment Approaches*

## 2.1 Introduction

The research base in biotechnology and bio-inspired technology has grown at a rapid rate, with major efforts being made by the United States (US), the United Kingdom (UK), the European Union (EU) and Asia. One indication of the importance of biotechnology in defence-related work can be seen by the significant investments that have been made in this area over the past decade, both in funding for research and development (R&D) and in the development of specific military programmes focused on biotechnology. A primary driver for this broad-based support has been numerous studies released by the US National Academy of Sciences,[1] Framework 6–7[2] and the US Defense Science Board,[3] suggesting that nano-/biotechnology-based research in the areas of environmental security, materials, sensors, autonomy and human performance has broad implications for future war-fighter protection and national defence. The emphasis on enhanced protection and avoidance of technological surprise has been supported by early demonstrations of sensors systems for biological agent detection (see Lim *et al.* for a review[4]) and has led to further work in areas that range from biomimetic multifunctional materials for energy absorption to bio-inspired adhesives or coatings.

Approaches to defence-related spending for R&D have changed significantly over the past 10–20 years. To a large extent government research funding in the 60 years following World War II was dominated by Cold War priorities, particularly in the US.[5] For bio-related work, this translated into a focus on protection from biological warfare attacks, with most of the funding directed towards vaccine development, diagnostics and therapeutics. With the collapse of the former Soviet Union, however, the emphasis largely shifted as perceived

Bio-inspired Materials and Sensing Systems
By Peter Biggins, John Hiltz and Anne Kusterbeck

Published by the Royal Society of Chemistry, www.rsc.org

threat levels decreased and expenditures on bio-defence began to drop. Now, however, the global concerns over terrorism and uncertainty about the nature of the enemy have shifted attention once again towards protection from biological agent release. In addition, rapid breakthroughs in molecular biology, including sequencing of the human genome, development of DNA/RNA arrays and the ready availability of commercial systems for rapid sequencing or synthesis of DNA/RNA/proteins, have increased the potential for biological warfare. Weapons of Mass Destruction (WMD) can enable even small, non-technologically advanced countries to use bio-warfare agents since biotechnology has allowed easier production/weaponization of virulent agents and technologies are widely available.[1] Concerns over preparedness and avoidance of technological surprise have therefore led to steady or increased support for R&D in these areas by the developed countries.

## 2.2 Effect of Globalization on Investment

Perhaps nothing better defines the current state of R&D funding for defence biotechnology than the concept of globalization. The widespread, rapid availability of knowledge made possible by information technology and electronic media,[6] the increased emphasis on dual-use technologies, the massive worldwide efforts made for the Human Genome Project and the changing nature of war itself have all contributed to this trend. In addition, enhanced global trade and economic interdependence of large corporations have played a role in support for science and technology, while at the same time heightening concerns about national autonomy, security and inequality, especially with respect to the largest investors, China, India and the US.[7] Finally, the interface of nanotechnology with biotechnology (or nanobiotechnology) is creating significant opportunities worldwide for new collaborations and multinational programmes designed to exploit this field.[1,8]

What is the impact of this globalization on biotechnological investment? Along with government funding, increased commercialization of defence-related innovations has led to spin-off companies that directly exploit military R&D. There is a growing emphasis on technology transfer and large biotechnology companies that operate in many different countries have also accelerated development of biotechnologies that have both civilian and military applications. Despite this trend, significant differences remain in investments by individual countries, as discussed in the following sections.

## 2.3 Overview of Investment by Key Countries

Although almost every developed country has funding directed toward defence R&D, only a few are major contributors to the biotechnology R&D area. As illustrated by Table 2.1, the funding for research in general and the specific level of support provided by the various government agencies varies greatly from country to country, with the overwhelming majority of the money being spent by the US. Other relative measures of spending, for example the R&D budget

**Table 2.1** Gross Domestic Expenditure on R&D (GERD).

| *Country* | *% of GDP* | *Government Investment (% of total)* | *Industry Investment (% of total)* | *Government Budget Appropriations for R&D, 2008* | |
|---|---|---|---|---|---|
| | | | | *% of GDP* | *% of which is defence R&D* |
| Australia | 2.01 | 38.37 | 57.23 | 0.46 | 7.70 |
| Canada | 1.88 | 31.42 | 49.40 | 0.57 | 4.00 |
| France | 2.08 | 38.42 | 52.44 | 0.75 | 27.70 |
| Germany | 2.54 | 27.76 | 68.07 | 0.79 | 6.00 |
| Italy | 1.13 | 48.32 | 40.42 | 0.64 | 1.20 |
| Japan | 3.44 | 15.63 | 77.71 | 0.70 | 5.20 |
| Korea | 3.47 | 24.80 | 73.65 | 0.96 | 18.00 |
| United Kingdom | 1.79 | 29.33 | 47.19 | 0.67 | 24.20 |
| United States | 2.68 | 27.73 | 66.44 | 1.00 | 56.60 |
| G7 | 2.53 | 26.64 | 65.80 | 0.87 | 38.60 |
| EU15 | 1.90 | 33.36 | 55.60 | 0.74 | 13.10 |
| EU27 | 1.77 | 34.11 | 54.98 | 0.69 | 12.50 |
| China | 1.49 | 24.62 | 70.37 | | |
| India | 0.71 | 80.81 | 16.11 | | |
| Russian Federation | 1.12 | 62.62 | 29.45 | 0.40 | 0.52 |

Source: Organisation for Economic Co-operation and Development (OECD).

as the percentage of GDP and the 56% level of the R&D funding in the US spent on defence-related research, further confirm the high priority placed on this area by the US. Recent moves by the EU to institute a central funding organization similar to the US National Science Foundation (NSF)[9] and significant changes taking place in government support of science and technology by India and China[6,10] have also heavily influenced R&D expenditure patterns. A comprehensive 2009 report by R&D Magazine and Battelle points out that, although the dominance of the US in innovation, intellectual property and overall spending is likely to continue in the short term, the gains in China and India are rapidly moving these countries ahead, with the second highest spender, Japan, starting to fall behind.[11] Figure 2.1 provides an overall picture of where each country stands currently with respect to relative spending on R&D.

Individual countries have also applied different strategies towards developing applications with regard to defence and biotechnology. The key nations investing in this area are highlighted below.

### 2.3.1 United States

As discussed above, the leader and dominant investor in military R&D is the US, which also has made one of the largest and earliest investments in biotechnology for defence. In the past, bio-focused research by US military/DoD was based largely on the requirement for chemical, biological, radiological, nuclear and explosives (CBRNE) defence. A major network of medical/vaccine related interest groups as well as biothreat agent research at dedicated facilities

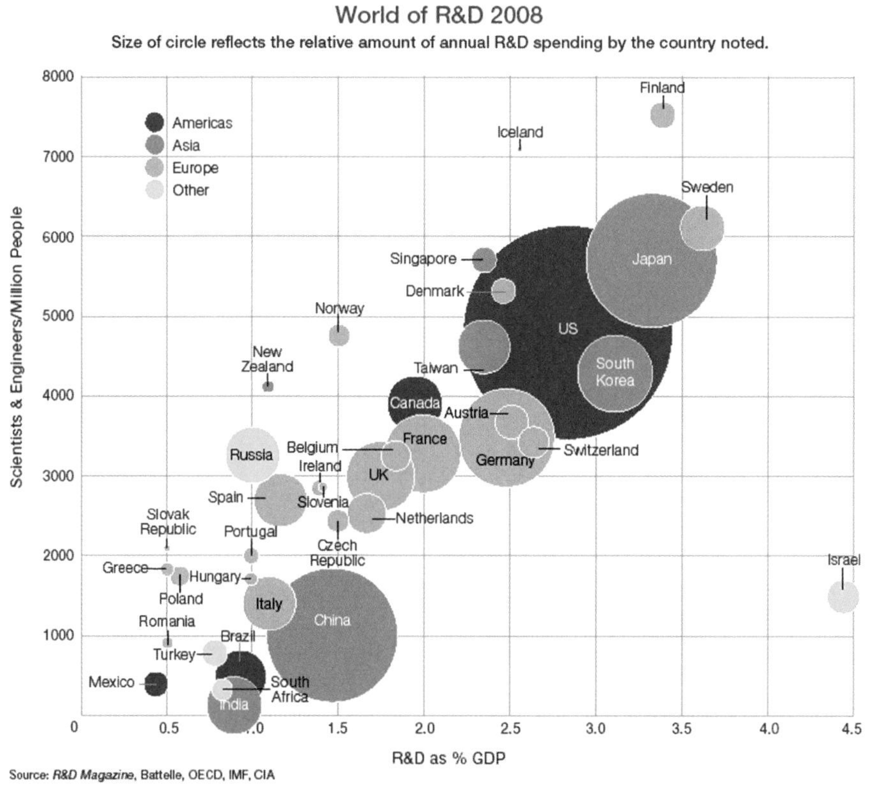
World of R&D 2008
Size of circle reflects the relative amount of annual R&D spending by the country noted.
Americas
Asia
Europe
Other
Scientists & Engineers/Million People
R&D as % GDP
Finland
Iceland
Sweden
Japan
Singapore
Denmark
US
Norway
New Zealand
Taiwan
South Korea
Canada
Austria
France
Germany
Switzerland
Russia
Belgium
Ireland
UK
Spain
Slovenia
Netherlands
Slovak Republic
Portugal
Czech Republic
Israel
Greece
Hungary
Poland
Italy
China
Romania
Brazil
Turkey
Mexico
South Africa
India
Source: R&D Magazine, Battelle, OECD, IMF, CIA

**Figure 2.1** Chart illustrating relative amounts of R&D spending by different countries as a function of % GDP and proportion of scientists and engineers in the population.

for bio-warfare defence, such as the US Army Institute for Infectious Diseases (USAMRIID) built up over the years with clear delineations and limitations on who could be supported. As the biotechnology 'revolution' and advances in molecular biology took place in the 1990s, however, the attention of the defence community shifted and the boundaries became less clearly defined. Though CBNRE areas continue to have a high priority in US defence research, the focus has also broadened to such areas as human performance enhancement, smart materials, nanobiotechnology and bio-inspired computing. This was based in large part on a comprehensive National Research Council 2001 study (begun long before 9/11) in which the authors recommended specific investments and priorities for research funding in biotechnology by the DoD.[3] As the post-9/11 security concerns of bioterrorism increased, funding for bio-related work also increased across all US agencies and many of the previous barriers were broken down. For example, the rapid establishment of the Department of Homeland Security (DHS) moved the Department of Energy (DoE) national labs into an umbrella organization with a specific US national security focus, resulting in the creation of biotechnology/biosecurity R&D programmes by ten leading national laboratories previously focused almost exclusively on nuclear research. These included Sandia National Laboratory, Pacific Northwest National Laboratory and Oak Ridge National Laboratory, research programmes directed towards energy, materials, bio-defence and homeland security.

A later US Defense Science Board study from 2006[12] concluded that the capabilities required to protect against surprise were most likely to be provided by biotechnology, nanotechnology and information technology and recommended increased funding in these areas by the DoD. Accordingly, DoD laboratories (Air Force research laboratory (AFRL), Naval Research Laboratory (NRL) and the Army Research Laboratory (ARL)), services and agencies (Defense Advanced Research Projects agency (DARPA) and Defense Threat reduction Agency (DTRA)) have established major initiatives in biotechnology for defence to develop technologies for war-fighter protection, the Global War on Terror (GWoT) and homeland security. Figure 2.2 provides an overview of all the DoD organizations actively involved in some aspect of R&D in this area, funded in fiscal year 09 at almost $80 billion. Going beyond the biothreat boundaries, bio-inspired approaches for biomaterials are specifically listed as strategic areas of research for the DoD,[12] with substantial overall funding for R&D related to biotechnology or bio-inspired approaches. As the largest service with a lead role in CBRNE, the Army has established the Institute for Collaborative Biotechnology (ICB), which received $44 million during the first five-year period (2003–2008) to pursue biomolecular sensors, bio-inspired materials and energy, bio-discovery tools, bio-inspired network science and cognitive neuroscience. The Army's budget for ICB in the second five years, fiscal years 2009 to 2013, totals $84 million (http://www.icb.ucsb.edu/).

The US AFRL has also gone so far as to include these bio-based approaches as one of the 'game changers' for future defence applications. The AFRL believes that research in nanoscale and biologically inspired materials may provide revolutionary sensors, self-assembling and self-healing structures for next-generation war-fighter capability as well as ensure that AFRL leads the nation in exploiting cutting-edge research and technology transition in nano/biomaterial for Air Force and DoD application.[13]

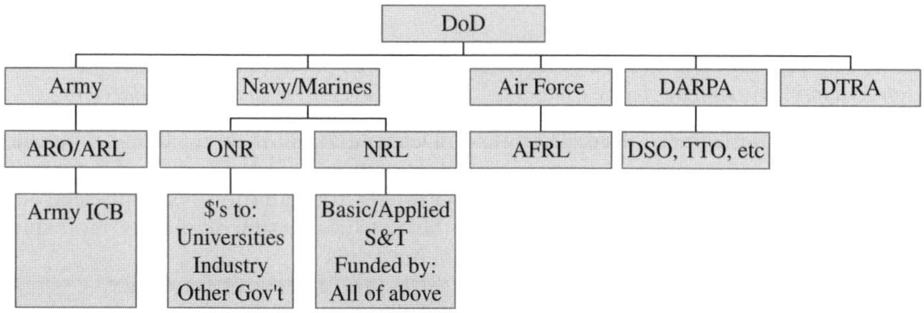
DoD
Army
Navy/Marines
Air Force
DARPA
DTRA
ARO/ARL
ONR
NRL
AFRL
DSO, TTO, etc
Army ICB
$'s to:
Universities
Industry
Other Gov't
Basic/Applied
S&T
Funded by:
All of above

**Figure 2.2** Diagram of principal organizations within the US Department of Defense involved in R&D related to biotechnology.

The DoD's, DARPA and DTRA have undergone similar shifts in research support to investigate biotechnology for defence (Table 2.2). Under its strategic thrust areas, DARPA has initiated programmes related to biologically inspired approaches to material synthesis, platforms or systems, as well as real-time modulation of protein conformation as a way to design sensors for engineered targets (DARPA, http://www.darpa.mil). DTRA, always a bio-defence entity, has seen substantial funding increases to support its mission of protecting the war-fighter by developing new biosensors and broad spectrum countermeasures against chemical and biological attacks.

Finally, the DoD and DHS have not been the only government groups to focus on biotechnology. A similar paradigm shift has occurred in the NIH. Previously funding primarily basic academic research in the life sciences, the changing model for NIH/industry support begun in the late 1990s has continued so that defence-related research in the historically basic/non-military US institution has increased. Recently (2008) biosecurity was added to the mission and strategic vision of the NIH (www.nih.gov).

### 2.3.2 United Kingdom

The UK Ministry of Defence's Defence Science and Technology Laboratory (Dstl) leads the science and technology (S&T) sector's response to the current and future defence needs. The research base and support for defence-related research in the UK is largely drawn from the academic and industry sectors and international partners. The UK Defence Technology Plan recognizes biotechnology and bio-inspired research explicitly as an important emerging technology area which will require investment (www.science.mod.uk/strategy/dtplan/default.aspx). Biotechnology-related research is principally funded through the Biotechnology and Biological Sciences Research Council (BBRSC) (www.bbsrc.ac.uk/home/home.aspx), and the Engineering and Physical Sciences Research Council (EPSRC) (www.epsrc.ac.uk/about/facts/Pages/mission.aspx), the UK government research councils responsible for basic research in non-medical sciences. As seen clearly in the mission and research priorities of the BBSRC, the work they fund is related to the fundamental understanding of biological systems rather than specific technologies, though such topic areas as Bioenergy, Global

**Table 2.2** Recommended investment areas in biotechnology for the US Army.

| *High Priority Investment Areas* | *Basic Research Areas with Barriers* |
|---|---|
| 3-D (volumetric) memory for rugged data storage | Determination of target threat molecules for sensors |
| Self-replicating systems for wound healing | Proteins for radiation-resistant electronics |
| Small-scale vaccine production | Hierarchical design models for bioinspired materials |
| Shock therapeutics | Structural interfaces for device substructures |
| Vaccine stratification by genomics | |

Source: Opportunities in Biotechnology (2001), National Academy Press.

Security, Nanoscience through Engineering to Application, Bionanotechnology and Systems Approach to Biological Research and Synthetic Biology might all be considered to have relevance to defence needs and requirements. The newly published 2010–2015 Strategic Plan outlines the importance of biotechnology for health, medicine and food security, as well as emphasizing the expansion of multidisciplinary research. Some of the work carried out by the BBSRC and EPSRC research is applicable to the defence and security research requirements for novel sensing and materials solutions. Despite the world-class level of research and innovation that continues in the UK, the funding picture for the government funding of R&D in the next few years is less clear.

### 2.3.3 European Union

As stated by the European Union (EU) 7th Framework Programme, funding support in the EU encompasses R&D efforts focused on addressing security, political stability and other world events that threaten the EU. The EU has recognized its unique vulnerability to these threats due to the fluid state borders, highly mobile population and proximity to countries with volatile or unstable governments and has concluded that technology plays a key role in addressing the new security challenges in this field (see 7th Framework Programme website, http://cordis.europa.eu/fp7/home_en.html). The EU has therefore taken steps to strengthen its research base accordingly with the stated goal to make Europe the leading world forum for science and technology by supporting work in nine thematic areas, including five that could be seen as being linked to military protection, *i.e.* 1) health, 2) food, agriculture and biotechnology, 3) nanosciences, nanotechnologies, materials and new production technologies, 4) energy and 5) environment (including climate change). One outcome of the vision to make the EU a highly competitive player in the global research community was the establishment in 2007 of the independent European Research Council (ERC). Modelled on the US NIH and DARPA, the ERC uses well-recognized scientists and a peer-reviewed process of proposals to fund selected high-risk, high-payoff basic research projects from across the EU countries. In the first round of funding, 34% of the projects were from the life sciences, examining such topics as biomolecular computers and silks as biomimetic polymers.[9] For the future, most indicators suggest that Europe will lose economic ground mainly because of lower population growth and dependency on the rest of the world, in particular for energy.[11]

### 2.3.4 China

Though China placed little emphasis on science and technology for much of the last century, the country is making rapid progress in its goal to overtake the US as the global leader in R&D. Following years of a low-technology approach to defence, China has become a 'fast follower' on almost every aspect of research, especially in the area of information technology.[6,8] The best source for data on China's investment approaches is the Outline of National Medium- and

Long-term Science and Technology Development Plan (2006–2020), released in 2006. The Outline identifies innovation as a new national strategy and states that the goal of this plan is to put China into the rank of innovative countries by 2020. To fulfil the mission the country will invest more than 2.5% of the GDP in R&D, with contributions of S&T to economic development exceeding 60% and a reduction of dependence on foreign technologies to below 30%.[14] Though China has significant catching up to do, Erickson and Walsh[6] have suggested that, cultural differences notwithstanding, when it comes to military modernization, current US and Chinese strategies for defence-related R&D increasingly share key characteristics that include a growing reliance on commercial-sector R&D, increasing emphasis on cooperative global research ventures and continued state-funded defence research programmes for developing next-generation technologies.[6] One example of the change is a shift in the major contributors to defence-focused R&D over the past ten years. In that time, the major source of funding has gone from China's government and military to China's non-public commercial sectors, which fund approximately 60–70% of the national R&D effort. The 2.5% of GDP goal outlined in the MIST plan is still in the future (the 2008 level was 1.49%), however, in terms of national R&D spending, China continues to advance. Though the exact amount of defence R&D spending is not readily available, it is most likely on a par with China's defence spending, which has averaged increases of 15% annually since 1990,[15] and which shows no sign of slowing.

### 2.3.5 India

Over the past decade, India, with an increasing GDP and the growth of its middle class, has been reported to be the rising star in science and technology, based in large part on the success of the country in the IT arena.[16] Despite these gains, the Indian government investment in R&D is the lowest in the world for a developing nation (approximately 0.8% of GDP), though the goal is to have it rise to 2%. Investment decisions for the government are driven largely by the National Committee of Science and Technology (NCST), as well as the Science Advisory Council, which created a Department of Biotechnology in the 1980s. Almost 90% of the R&D money comes from the government, with relatively little private sector contribution. Historically the state support has gone to military applications,[7] however several prominent features of the S&T landscape in India have changed with the migration of multinational companies and the establishment of private R&D efforts in biotechnology and pharmaceuticals. Despite the influx of funding for these sectors, defence-focused investments are not focused or seen as a high priority by the government.[16]

### 2.3.6 Japan

For a long time Japan was the second-highest investor in R&D, but the country is now undergoing a change as the economy stumbles and its debt levels rise.

Nevertheless, government-supported research remains strong and is expected to continue for the foreseeable future at the 3.4% GDP level of previous years.[11] A relative newcomer to formal funding agencies, the Japan Science and Technology Agency was established in 2003 to guide research in basic sciences.

### 2.3.7 Russia

While Russia has held on to its ranking as the fourth largest researcher community in the world, behind the US, China and Japan, its proportional spending on R&D has fallen to just over 1% of the GDP, with government support accounting for almost 70% of that figure. New government programmes have also been established to support R&D in priority industries, including space and aviation, nanotechnology, biotechnology and software.[11]

## 2.4 Future Trends

Although no one can accurately predict future investments in R&D, there are several significant trends emerging. The continued globalization and rapid spread of technologies, with the increased use of information technology and autonomous systems on the battlefield, is already leading to heightened awareness of the important role bio-related discoveries and innovations can play in defence and security. As discussed in previous sections, countries are continuing to provide resources and funding to support biotechnology as well as developing specific strategies to develop defence related bio-inspired technologies for the war-fighter, especially in the US. Finally many experts have suggested that by 2025 an even more interconnected world will see the rise of China as the second global economy.

## 2.5 Conclusion

Defence research organizations will need to continue adapting to the changing nature of national security, terrorist threats and asymmetric warfare, including the possible use of unknown chemical or biological agents. In addition to globalization, the fluid nature of borders and the vast amount of knowledge made readily available to anyone by IT systems make it critical that governments invest in R&D to avoid surprise and prevent disruptive technologies. The internet operates seamlessly and without borders or control. From the perspective of bio-inspired research, sensing technologies, biomaterials and bioenergy production could all be beneficial in meeting the challenges within the defence, security and civilian sectors. Other related areas including nanotechnology, health/medical and performance enhancement will also continue to drive investments.

# References


1. NRC Committee on Defense Intelligence Agency Technology Forecasts and Reviews, *Avoiding Surprise in an Era of Global Technology Advances*, http://www.nap.edu/catalog/11286.html, National Research Council, 2005.
2. M. Farre, L. Kantiani, S. Perez and D. Barcelo, *TrAC: Trends in Analytical Chemistry*, 2009, **28**, 170–185.
3. NRC Committee on Opportunities in Biotechnology for Future Army Applications, *Opportunities in Biotechnology for Future Army Applications*, The National Academies Press, 2001.
4. D. V. Lim, J. M. Simpson, E. A. Kearns and M. F. Kramer, *Clin. Microbiol. Rev.*, 2005, **18**, 583.
5. J. P. Dunne and D. Braddon, *Economic impact of military R&D*, Flemish Peace Institute, 2008.
6. A. S. Erickson and K. A. Walsh, *Tech. Soc.*, 2008, **30**, 349–361.
7. A. Segal, *Tech. Soc.*, 2008, **30**, 423–428.
8. E. S. Michelson, *Tech. Soc.*, 2008, **30**, 405–410.
9. M. Antonoyiannakis, J. Hemmelskamp and F. Kafatos, *Cell*, 2009, **136**, 805–809.
10. R. W. Nichols, *Tech. Soc.*, 2008, **30**, 437–450.
11. M. Grueber and T. Studt, in *R&D Magazine*, Advantage Business Media, Rockaway, NJ, 2009, pp. 20–34.
12. D. S. Board (ed.), US Defense Science Board, 2007, *2006 Summer Study on 21st Century Strategic Technology Vectors Volume II: Critical Capabilities and Enabling Techologies*, Office of the Under Secretary of Defense for Acquisition, Technology, and Logistics Washington, D. C.
13. Air Force Research Laboratory, *Air Force Capabilities Based Science and Technology Strategy 2030: AFRL's Corporate Long Term Strategic Plan*, 2007.
14. J. T. Ratchford, *Tech. Soc.*, 2008, **30**, 211–233.
15. S. Jian, *Tech. Soc.*, 2008, **30**, 235–241.
16. R. Narasimha, *Tech. Soc.*, 2008, **30**, 330–338.

CHAPTER 3

# Conceptual Approach

## 3.1 Introduction

The scope for bio-inspired applications to sensors and sensing systems is only limited by the imagination; early prototype systems will have to be practical but once established could be developed in the same way current technology has developed over time. Flight is an example of a highly desirable capability that has been realized by the continuing development of aircraft. If the history of the development of aircraft, which was inspired by man's desire to fly (bio-inspiration) is considered, it required that certain technological barriers be overcome and the growth of the engineering/industrial base to provide the key tools required for component optimization and performance.

## 3.2 Operational Requirements and Concepts of Operation

The type of sensor or sensor system developed will depend on how it will be deployed and the required level of functionality. For instance, a passive sensor would have no need for power. It could be interrogated from a distance to obtain data. Alternatively, the system might be sedentary and have an engineered suite of sampling, transduction and processing capabilities that have minimal power and energy requirements that could be met by harvesting energy from the environment. Or the system might have a requirement to be highly mobile and carry a programmable range of sensors, or be networked to other sensors or sensor systems to provide enhanced area coverage. The power requirements for this system would be considerably greater.

Bio-inspired Materials and Sensing Systems
By Peter Biggins, John Hiltz and Anne Kusterbeck

Published by the Royal Society of Chemistry, www.rsc.org

**Table 3.1** Factors affecting the design and selection of sensors and sensing systems.

| | *Covert* | → | *Overt* |
|---|---|---|---|
| Deployment | Static, scatterable<br>Interrogate from a distance | Sedentary<br>Unmanned<br>Single or part of network | Mobile<br>Part of a network |
| Function | Passive sampling | Active sampling characterized trans-duction systems processing | Active sampling |
| | Functionalized sur-face using holo-graphic structures, quantum dots and beads | | Directional |
| | | | Programmable range of sensing systems<br>Significant processing |
| Material | Biodegradable | Sacrificial/degradable retrievable | Durable |
| Power | None | Minimal | Significant power |
| Asset value | Low | Medium | High |

Illustrated in Table 3.1 are examples of how the range of applications of sensors and sensing systems may be employed in various operational scenarios. This range is broad and has the potential to address a wide range of requirements.

The types of materials used in the sensor would depend on how it is deployed. If there is a requirement that the sensor be covert, then this could impact on its size or require the use of camouflage. If the sensor is small and inexpensive then it may be desirable to use biodegradable materials, whereas if the sensor or sensor system is an expensive asset, the materials used in its construction might be more robust and have improved environmental resistance. As the size and complexity of a sensor or sensor system increases, the integration of the components of a sensor or the component sensors of a system may become critical. Regardless of the match between requirements and operational scenarios, the cost of the technology must also be considered and may ultimately be the deciding factor in choosing the technology that will be used.

## 3.3 Conceptual Goal

One approach to facilitate the application of biotechnology and bio-inspiration to sensors and sensing systems is to develop a conceptual goal.[1] In the broadest sense the conceptual goal may never be fully realized. However, the conceptual goal enables a vision to be developed that allows or assists in the identification

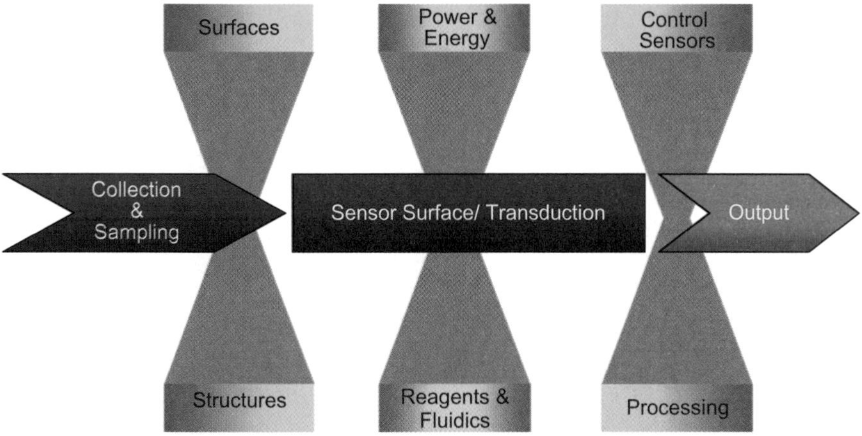
Surfaces
Power & Energy
Control Sensors
Collection & Sampling
Sensor Surface/ Transduction
Output
Structures
Reagents & Fluidics
Processing

**Figure 3.1** Schematic showing the key elements and enabling technologies for an autonomous sensing system.

of enabling technologies that are required or need to be developed for the realization of the vision.

The Situational Awareness Sensing System (SASS) that can operate both autonomously and/or in a distributed fashion is a conceptual goal for a sensor or sensing system. A schematic of a SASS is shown in Figure 3.1. The value of this approach is that it allows the discrete building blocks to be explored and optimized based on the current and emerging science and technology base. In addition, this model illustrates the importance of how components must be integrated to achieve the desired level of functionality. Biological systems do this with a level of ease that we cannot readily replicate at this time.

## 3.4 Enabling Technologies

The key elements of a SASS sensing system are the ability to collect from or sample the environment, an interface with a sensor surface and corresponding transduction system that produces a signal for processing and the production of an output. For such a system to be built requires collection and sampling surfaces and mechanisms, appropriate structures and materials, fluidics and reagents if required to process the samples, a transduction system to identify or sense the analyte, a control system and the ability to process and communicate results. The SASS model could be used as a guide to identify the science and technology required to develop new sensing systems for defence and security applications or as a guide to evolutionary change of both new and existing systems.

While a system view provides an overall framework, the component analysis (Figure 3.2) identifies the specific areas of science and technology where developments are required for the realization of a SASS. By focusing on each component separately, workable projects that can be optimized and inserted into the SASS concept can be identified.

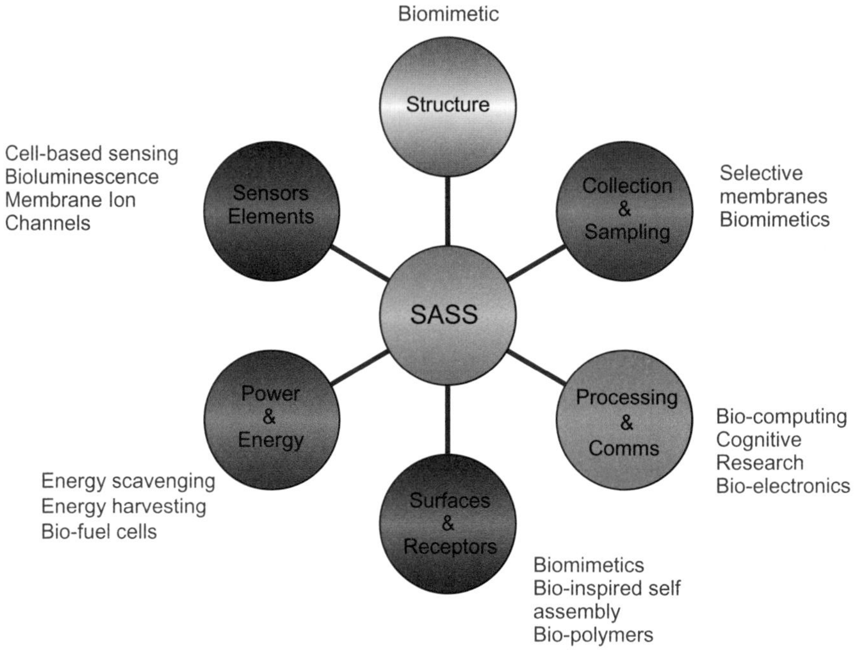
Biomimetic
Structure
Cell-based sensing
Bioluminescence
Membrane Ion
Channels
Sensors
Elements
Collection
&
Sampling
Selective
membranes
Biomimetics
SASS
Power
&
Energy
Processing
&
Comms
Bio-computing
Cognitive
Research
Bio-electronics
Energy scavenging
Energy harvesting
Bio-fuel cells
Surfaces
&
Receptors
Biomimetics
Bio-inspired self
assembly
Bio-polymers

**Figure 3.2** The successful realization of the SASS concept requires a number of enabling technologies. These include structure, surfaces and receptors, collection and sampling, sensing/transduction and power and energy.

A short overview of enabling technologies is discussed below, with a more comprehensive examination for each in subsequent chapters.

### 3.4.1 Collection and Sampling

Collection and sampling encompass the most critical aspect of detection, *i.e.* to be detected a sample of sufficient concentration must be collected and delivered to the transduction element. Many sampling techniques and technologies have been developed, for instance, techniques for entraining aerosols into liquid streams for subsequent concentration and analysis;[2,3] however, the devices used for sampling are cumbersome, require significant user interface and are inefficient at collecting trace materials. The study of natural systems and bio-inspiration has led and will lead to novel approaches to sampling and collection.

Improvements in sampling and collection of samples are often neglected in favour of developing elegant transduction systems. This may be acceptable within a laboratory environment but can result in problems for deployed sensing systems from the size, cost and power perspectives. However, if more efficient sampling and collection systems can be developed for deployed systems, it may be possible to simplify the transduction elements.

### 3.4.2 Structures

Structure, from the materials' perspective, involves many aspects of the SASS system including collection and sampling, receptors, transduction, robustness and environmental resistance. Structure also includes the way in which the components of the system are integrated, for instance, the structure of the 'sensing train', from collection of a sample to the signalling event. Structures found in Nature are often multifunctional and the combination of functions is inspiration for reducing the number of components in a SASS. For instance the surface of a deployed sensor might not only be optimized for collection and sampling but also incorporate environmental resistance and stealth characteristics (camouflage).

### 3.4.3 Receptors and Surfaces

Surfaces and receptors are in essence an extension of the structure of a material, that is, the properties of surfaces and receptors are directly related to their structure down to the nanometre level. This is an area where much has been learned from Nature. There is a large number of existing bioreceptors including antibodies, enzymes, olfactory binding proteins, DNA/RNA probes, synthetic ligands and cell surface receptors[4–6] that have evolved with the sole purpose of binding to target ligands involved in specific biological or biochemical processes. Many of these reagents find '*in vitro*' applications in various sensors and detection systems. In addition, both biomimetic and bio-inspired approaches to surfaces and receptors have led to development of sensors (or components of sensors) with improved sensitivity and specificity for analytes of interest. Approaches such as the development of molecularly imprinted polymers are bio-inspired and are directed at addressing some of the shortcomings of biological (natural) systems including temperature, stability, pH variations and organic solvent sensitivity.

### 3.4.4 Sensing and Transduction

Considerable research and development has been focused on the development of a range of transduction systems. These systems include those based on optical, electrochemical, acoustic, piezoelectric, mass spectrometric and magnetic transduction and detection.[6,7] There is much to be learned from Nature about sensing and transduction. For instance, some insects and reptiles are able to detect heat or infra-red radiation for identifying prey and/or avoiding predators.

### 3.4.5 Processing and Communication

An important part for any sensing system is how information is processed. Smart processing algorithms, such as probabilistic networks, have been developed for many detection systems. It has been suggested that bio-inspired molecular computers that are capable of massively parallel computing may circumvent some of the shortcomings of traditional silicon based computers. However, this will require the development of algorithms that can take full advantage of the promise of massively parallel computing found in Nature. There is also much to be learnt from Nature concerning communication. Many systems found in Nature, although data transfer is slow, have great specificity and therefore are highly secure. The use of pheromones by insects is an example of this type of communication.

### 3.4.6 Power and Energy

A key aspect to the deployment of any sensing system is the incorporation of an appropriate power source. In most instances for deployed systems, power is supplied by batteries. There are obvious logistical issues related to the provision of power using batteries. One is getting the batteries to and from the site where they are needed. If the sensor is remote and stealthy, it may not be possible to change batteries. Batteries have limited lifetimes and are also heavy. For deployed soldiers they may add significantly to the weight of kit they must carry. Bio-inspired systems will find applications to a number of common problems facing defence and security forces in the area of power and energy. The development of self-sustaining systems is a truly bio-inspired approach. Researchers are looking at ways in which energy can be harvested from the environment. For example, estuarine and marine sediments can be used to power unattended sensing systems in rivers and oceans.[8] Other research groups are examining how to derive power from tree saps and mammalian fluids. Novel approaches to harvesting or scavenging power from the environment offer the potential to deploy new types of sensing packages that use what is available in the environment to provide power for remote detection, communication and motion.

## 3.5 A Larger Vision of the SASS Concept

A larger vision of the SASS concept is shown schematically in Figure 3.3. It has a hierarchical structure similar to many found in natural systems. This structure exists from the subcomponent level (structure, sampling, receptors, sensing, processing and power) through the component level to the system and system of systems level. The SASS concept is applicable not only to a single entity but also as part of a system, or even a system of systems in the same way that a single organism, a multi-cellular organism or group of organisms is configured in Nature.

At all levels there is much to be learnt from natural systems. Descriptive words such as multifunctional, integration, specificity, sensitivity, robustness, self-healing, self-sustaining and processing and communication are applicable at the subcomponent through the system of systems levels of the concept. Their importance will depend on the final design of the operational system. If one considers communication, it varies from the component level where it may be

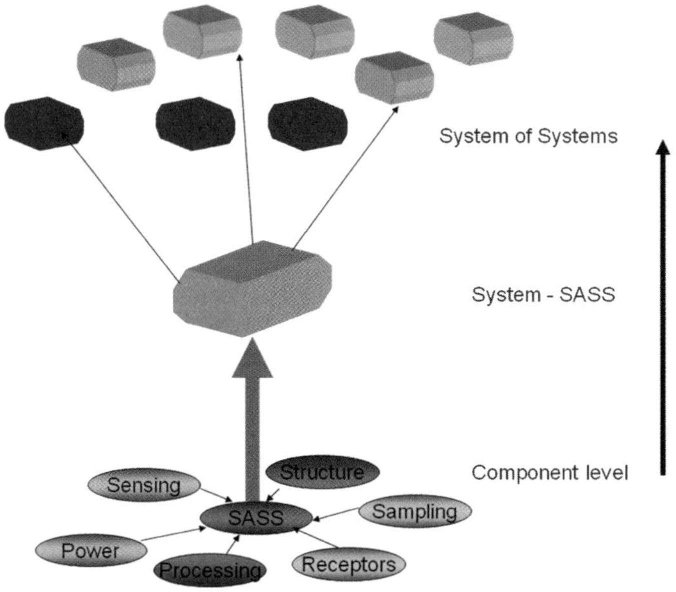
System of Systems
System - SASS
Component level
Sensing
Structure
SASS
Sampling
Power
Processing
Receptors

**Figure 3.3** A schematic representation of a larger vision of the SASS concept.

primarily internal communication to the system and system of systems level where it may be primarily external. Processing may also change significantly as one moves from the component to the system of systems level. At the component level it may involve a transduction event while at the system and system of systems level it may involve the processing of data from many transduction events or the fusing of data from a number of different sensors and decisions on possible courses of action.

## 3.6 Conclusion

The SASS conceptual approach provides a basis for exploitation and development of biotechnology and bio-inspiration for sensors and sensing systems from the subcomponent through the system of systems level. The following chapters will provide more details to the enabling technologies that can be used to realize the concept of a 'Situational Awareness Sensing System'. To achieve successful integration of bio-inspired technologies for the production of advanced systems that are superior to those currently in use will require an interdisciplinary approach. This will ensure that a stovepipe type of approach that does not consider the system as a whole is avoided.

## References

1. P. D. E. Biggins, A. Kusterbeck and J. A. Hiltz, *Analyst*, 2008, **133**, 563–570.

2. C. Cox and C. Wathes, *Bioaerosols Handbook*, Lewis Publishers, Boca Raton, FL, 1995.
3. *Sensor Systems for Biological Agent Attacks: Protecting Buildings and Military Bases*, National Academies Press, Washington, DC, 2005, Ch. 4, 5, 6, 7, 8 and 9.
4. D. V. Lim, J. M. Simpson, E. A. Kearns and M. F. Kramer, *Clin. Microbiol. Rev.*, 2005, **18**, 583.
5. J. J. Gooding, *Anal. Chim. Acta*, 2006, **559**, 137–151.
6. E. Kress-Rogers, (ed.), *Handbook of Biosensors and Electronic Noses: Medicine, Food, and the Environment*, CRC Press, Boca Raton, FL, 1996.
7. J. Schultz, M. Mrksich, S. N. Bhatia, D. J. Brady, A. J. Ricco, D. R. Walt and C. L. Wilkins, WTEC Panel Report on International R&D in Biosensing, Final report, World Technology Evaluation Center, Inc. (WTEC) NTIS, US Dept of Commerce, Baltimore, MD, August 2004.
8. L. M. Tender, C. E. Reimers, H. A. Stecher, D. E. Holmes, D. R. Bond, D. A. Lowy, K. Pilobello, S. J. Ferig and D. R. Lovley, *Nat. Biotechnol.*, 2002, **20**, 821–825.

CHAPTER 4

# *Structure*

## 4.1 Introduction

The structures of natural compounds, components or systems are characterized by their scale or size, their function, their morphology and the properties of the materials used in their preparation. This relationship is shown schematically in Figure 4.1 along with some of the factors that describe or define scale, function, morphology and materials.

In Nature structures exist on scales from the nanoscale ($10^{-9}$ m) through the microscale ($10^{-6}$ m) and the milliscale ($10^{-3}$ m) to the macroscale (m). Structures in Nature have an almost endless number of functions. These range from mobility, strength, toughness, camouflage and protection through sensing and self-repair. The morphologies of natural structures are responsible for properties such as hydrophobicity and camouflage. The materials and the way they are assembled or processed are critical to the structures found in Nature. The properties of the constituent materials used in natural structures influence the stability of a structure and its ability to survive and function in different environments. The conditions in these environments can include broad temperature and pH ranges, levels of moisture and humidity and the presence of toxins and predators. In addition, the starting materials used to build these structures must be available and low-temperature and low-energy pathways must exist for their assembly.

The interdependence of structure and properties in natural systems suggests that if one can understand the relationships between structure and scale, function, morphology and materials then this knowledge can be exploited to produce bio-inspired structures with similar properties. It also provides an opportunity to make structures with novel properties that are not found in

Bio-inspired Materials and Sensing Systems
By Peter Biggins, John Hiltz and Anne Kusterbeck

Published by the Royal Society of Chemistry, www.rsc.org

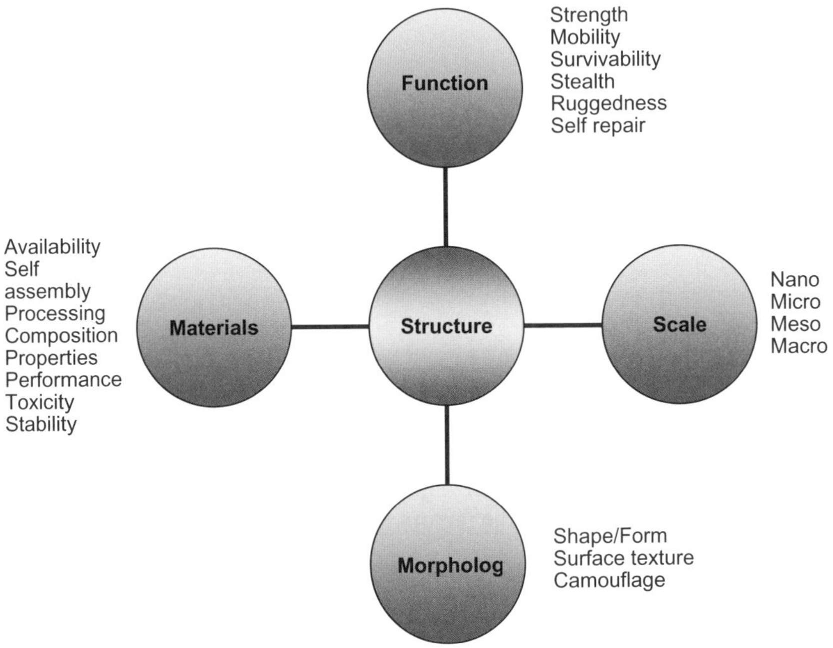
Function
Strength
Mobility
Survivability
Stealth
Ruggedness
Self repair
Availability
Self
assembly
Processing
Composition
Properties
Performance
Toxicity
Stability
Materials
Structure
Scale
Nano
Micro
Meso
Macro
Morpholog
Shape/Form
Surface texture
Camouflage

**Figure 4.1** Parameters affecting the structures and properties of natural materials and systems.

Nature. For example, these might include bio-inspired materials and structures that have resistance to environments that natural materials do not or materials that bind with specificity to synthetic compounds.

In this chapter several reoccurring themes associated with the structure of natural materials or systems will be discussed. These include hierarchical structures that are maintained from the smallest to the largest scale, a bottom-up approach to fabrication and multifunctional structures.

Structure will be discussed from the perspectives of scale, function, morphology and materials. When considering autonomous sensing and sensors, scale can be thought of in a number of ways. The first is at the component level. That is, how can a bio-inspired or biomimetics approach be used to prepare components of a system with the properties required for its intended application? The second is at the system level where the system is composed of a number of components. In this instance one might want to exploit the knowledge of how natural systems integrate the components of a system to optimize its function. These components might provide protection from the environment, the ability to sense the environment, the ability to react to the environment (mobility) and to provide the energy necessary to survive (function) in the environment. The third is at the system of systems level. Many natural systems (birds, ants, bees and fish for example) have developed behaviours (flocking, colonies, swarming and schooling) or social structures that provide protection or defence from predators, minimize energy use and promote their survival.

The approaches to the synthesis or fabrication of bio-inspired structures, some examples of bio-inspired structures and the major challenges in this area are also discussed.

## 4.2 Themes in Biological Systems

### 4.2.1 Hierarchical Structures

Natural materials and systems are characterized by hierarchical structures. Structure exists from the molecular scale through to the final scale of the material or system. Bone, a strong, lightweight and tough material, is an example of a material where structure is maintained from the molecular through the macro scale. The hierarchical structure of bone is shown schematically in Figure 4.2.[1]

The properties of bone are phenomenal when the materials used in its formation are considered. Bone is composed primarily of tropocollagen and hydroxyapatite. Tropocollagen is the molecular component of a collagen fibre and consists of three polypeptide chains coiled around each other to form a helix. The amino acids that make up the polypeptide chains are approximately 1 nm in length and the polypeptide chains that make up the tropocollagen helix are approximately 300 nm in length. A number of tropocollagen molecules associate to from collagen fibres and several of these associate to form collagen fibrils. There are distinct gaps within the packed collagen fibres. It is within

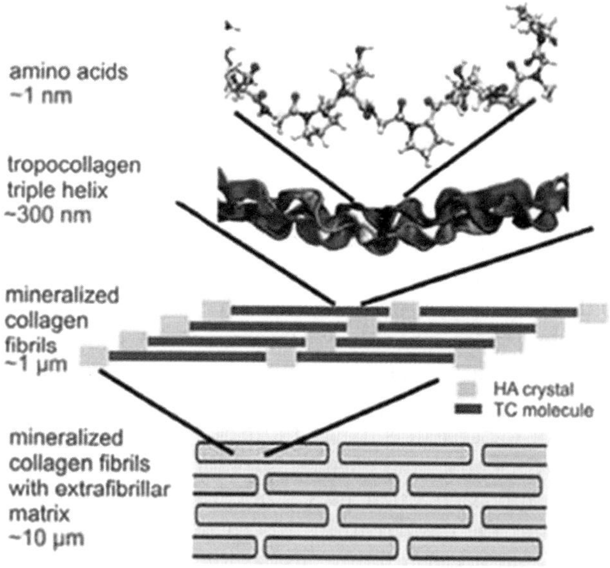
amino acids
~1 nm
tropocollagen
triple helix
~300 nm
mineralized
collagen
fibrils
~1 μm
HA crystal
TC molecule
mineralized
collagen fibrils
with extrafibrillar
matrix
~10 μm

**Figure 4.2** Schematic representation of the formation of bone from tropocollagen and hydroxyapatite. The hierarchical structure of bone from the nanometre through to the micrometre scale is shown. (M. J. Buehler, *Nanotechnology*, 2007, **18**, 295102, IOP Publishing Limited. Reproduced with permission.)

these gaps that mineral crystals form.[2] These plate-like crystals are made from hydroxyapatite and are packed into the collagen fibrils. The orientation of the mineral crystals is determined by the orientation of the collagen fibrils. Mineralized collagen fibrils are approximately 1000 nm (1 µm) in length. The collagen fibrils are subunits of lamella. The intra- and inter-lamellar collagen orientations differ. The orientation of intra-lamellar collagen fibrils is predominantly parallel. However, the orientation of inter-lamellar collagen fibrils can change by up to ninety degrees.

These structures are common to both cortical and trabecular bone found in humans and other mammals. Cortical bone is also referred to as dense bone and has between 5% and 10% porosity while trabecular bone has porosity ranging from 50% to 90%. The next level of structure in cortical bone is composed of osteons, which are between 100 and 300 µm in length. This is in contrast to trabecular bone whose next level of structure is composed of trabeculae that are between 75 and 200 µm in length.

This precise control of structure from the molecular level through to the macro level is responsible for the properties of bone. To mimic such a structure requires that synthetic pathways are available that can control structure from the molecular through the macro levels of scale.

### 4.2.2 Bottom-up *vs.* Top-down Approach to Fabrication

There are two approaches to fabricating structures: the bottom-up approach and the top-down approach.

In the bottom-up approach, the components are tied together to build larger and more complicated structures. The original components or structures then become subcomponents or subsystems of the larger system. The subcomponents and subsystems are linked, sometimes at many levels, until a complete system is formed. Although the starting components are small, they grow in complexity and finally become a complete system. This is the approach taken by Nature. The formation of bone described above is an example of the bottom-up approach. The chemical properties of constituent molecules and developing structures lead to formation of bone.

In the top-down approach, a system is broken down to gain insight into its compositional subsystems. This starts with an overview of the system that specifies, but does not detail, any first-level subsystems. Subsystems are then described in greater detail. This may require describing additional subsystem levels until the specification for the system is reduced to base elements. This is the approach most often used by scientists and engineers to fabricate bio-inspired structures.

The top-down approach often uses fabrication methods to cut, mill and shape materials into the desired form. Micro patterning techniques, such as photolithography and inkjet printing, also belong to this category. The drawback to this approach is that size is limited, for instance, by the wavelength of radiation used to fabricate the component or the detail required in the component. Examples of the successful use of a top-down approach to produce bio-inspired structures and materials will be discussed later in this chapter.

Attempts to copy Nature using the bottom-up approach are an area of great interest and involve concepts such as molecular self-assembly and molecular recognition. This approach has met with some success but it remains one of the greatest challenges to exploitation of the developing knowledge of the structures found in Nature. Advances in this area will enable the fabrication of a broad range of bio-inspired materials and systems.

### 4.2.3 Multifunctional Materials

Nature is rife with examples of structures and materials that have more than one function. In fact it might be said that biological structures are characterized by multifunctional materials. The wing of a butterfly is not only a part of the system responsible for mobility but also has colourful patterns that can, for instance, mimic the patterns of a poisonous butterfly and reduce attacks by predators. The toe pads of geckos are responsible for the adhesive forces that allow them to walk inverted on a surface and are also self-cleaning. This means that the adhesive qualities of the toe pads can enable or prevent adhesion. Plant surfaces have a number of functions. They act as an interface between the plant and the environment and in dry environments regulate the amount of radiation reaching the interior of the structure. In wet environments many plant surfaces are water repellent (hydrophobic). The shedding of water droplets leads to the removal of dust and pathogens from the plant surface. The shedding of water can also aid in plant respiration. Removal of moisture from the area of the stomata improves carbon dioxide diffusion into the plant.

## 4.3 Structural Parameters

### 4.3.1 Scale

Scale from the structural perspective is related to function and morphology. The hierarchical structure in Nature exists from the smallest component through to the final form of an organism or a constituent part. The existence of substructures down to the nanometre level requires that any discussion of 'structure' must consider the structure at the size scale that is responsible for its observed properties.

The selectivity of antibodies for antigens is related to both the size and the shape of these entities. The specificities of antibody/antigen reactions are based on binding interactions, such as hydrophobic, van der Waals and electrostatic interactions and hydrogen bonding. For these interactions to occur and hydrogen bonds to form, the surfaces of the antibody and antigen must have complementary shapes with respect to their surface structures. This maximizes affinity and ensures specificity.

The adhesive forces developed by the toe pads of geckos, insects and frogs are dependent on scale. The size of the ends of the hair-like structures that make up the toe pads determines the strength of the adhesive forces generated between the toe pads and the underlying surface. As the size of the ends decreases it has been observed that the strength of the attachment increases. This is the result of the improved contact between the toe pad structures and the underlying surface, which results from the toe pad end structures being smaller than the asperities on the surface.

The hydrophobicity of plant surfaces is also related to the size of the structures on the plant surface. Cone-like structures, micrometres in size, produce a surface roughness that prevents the wetting of the surface. In addition, the cones themselves are covered with nanometre-sized wax crystals that enhance the hydrophobicity of the surfaces.

Antibody/antigen interactions, toe pad adhesion and hydrophobicity are examples of the importance of scale or size of structures in the function of these structures. However, as was mentioned in the introduction to this chapter, scale is also a factor in the structure of systems and system of systems.

At the system level in Nature, scale is determined by a range of factors that have been optimized for the environment in which the natural system exists. These factors include mobility requirements, protection from predators and the natural environment, power and energy requirements and the ability to sense and react to the environment. Multifunctional materials, for instance, surface materials that provide protection, sensing and thermal-regulation capabilities, tend to reduce the overall scale (size and mass) of the natural system, reduce power requirements and improve mobility.

At the system of systems level, the scale or size of the group impacts on the individual system in a number of ways. For a school of fish these include defence, both from the perspective of the school being better able to detect a predator than an individual fish and from the greater chance of survival as a member of a larger group, an increased chance of finding food, an increased opportunity to reproduce and increased hydrodynamic efficiency.

### 4.3.2 Function

In her book *Biomimicry – Innovation Inspired by Nature*, Janine M. Benyus states that in Nature 'form follows function'.[3] The range of functions and consequently structures found in Nature and natural systems is seemingly endless. Mobility, vision, hearing, olfaction, camouflage, resistance to antigens and viruses, protection from predators and the environment, joining and adhesion, hydrophilicity, hydrophobicity, self-repair or healing, power generation and energy storage, sensor(y) input, processing and memory are just some of the functions of natural systems. The 'form follows function' assertion suggests that the structures of natural systems have evolved and will continue to evolve to optimize function or to adapt to changing environments.

The form that enables a particular function can vary significantly from one species to another. For instance, the structure of the system or systems responsible for mobility is dependent on the organism and the environment in which it lives. Humans and other bipeds walk on two feet using the muscles in the legs to pick up and put down feet. Other animals walk on four feet, birds and insects use wings to fly and fish control muscular motion to swim. In all instances the structures responsible for mobility have evolved to enable and often optimize movement in the environment, whether it is land, water or air, where the function is required.

The human hand is another example of the relationship between form or structure and function. The structure that enables the hand to grasp, hold and manipulate an object exists at many levels. The finger tips have a high density of nerve endings and are extremely sensitive to pressure and temperature. The skin at the fingertips is designed to aid in gripping and the musculature of the hand allows exertion of forces necessary to grasp and hold the object. The thumb opposing the fingers also facilitates the grasping, holding and manipulation of objects.

The forms found in Nature have evolved over extended periods of time (or through many generations). It follows then that if we want to use our knowledge of natural systems, from a biomimetic or bio-inspired perspective, an understanding of the relationships between the function and form in natural systems is an obvious starting place. That is, what is it in the form of a particular natural system that results in its function? Or how does the structure of a natural system lead to its observed properties? The understanding of the relationships between form and function can then be exploited to produce new materials and structures, systems or system of systems that these materials and structures comprise.

## 4.4 Biological Joining Technologies

Joining technologies, that is, technologies used to attach one material to another, are used every day. Adhesive bonding, welding, bolting and clamping are common examples of joining technologies. Some of these techniques result in very strong and quite permanent joints (welds and adhesives) while others are weaker and less permanent (bolts, buckles and clamps) and allow the joint to be disassembled and the joining technology to be reused. Nature has also developed joining technologies. Examples of bio-inspired joining technologies and how knowledge of the structures that enables them was exploited to make bio-inspired joints are discussed below.

### 4.4.1 Velcro

Velcro is one of the best known and most successful examples of a bio-inspired joining system. Georges de Mestral developed and eventually patented the material in the 1950s. He had come up with the idea in 1941 after investigating how burdocks (*Arctium lappa*) adhered to clothing and animal fur. He observed that the burdock had a number of hooks that would catch in loops in fabrics or tangles in animal fur. In the following years he worked to develop the new joining system that he called Velcro after the French word velours, meaning 'velvet', and crochet, or 'hook'. A micrograph of the hook-like structures on the external surface of a burdock is shown in Figure 4.3 along with the hook and

**Figure 4.3** From left to right, the hook structure on the surface of a burdock (*Arctium lappa*), the hook structure of Velcro and the loop structure of Velcro.

loops structures of the Velcro joining system. The system provides a joint with moderate strength that can be separated and reformed a number of times.

## 4.4.2 Toe Pad Adhesion

Dry and wet adhesion arising from toe pad structures of geckos, frogs and insects has been extensively studied. An excellent review of toe pad adhesion is found in the June 2007 issue of the MRS Bulletin.[4] The knowledge of the relationship between the structures responsible for wet and dry adhesion and function has been exploited to produce bio-inspired adhesive systems.

### *4.4.2.1 Dry Adhesion*

How do insects and geckos walk up walls and across ceilings? Research has shown that the bottoms of their feet have a large number of hairs called setae. The setae are further subdivided into structures called spatulae. It is the spatulae that are responsible for the observed adhesive properties of the toes and other surfaces of these species.

The ends of the setae of a beetle, a fly, a spider and a gecko are shown in Figure 4.4. All are shown at the same magnification. It is obvious from the figure that the spatulae become smaller in going from the beetle ($\sim 10\,\mu m$) to the fly ($\sim 2\,\mu m$) to the spider ($< 1\,\mu m$) and finally to the gecko ($\sim 0.2\,\mu m$). It has been shown that the strength of adhesion increases as the sizes of the spatulae decrease. This is due to the shape and size of the individual spatulae which allow the development of van der Waals forces between the spatula surface and the substrate.[5] The structures shown in Figure 4.4 are associated with what is termed dry adhesion.

The understanding of how the structure of the toe pad results in adhesion has led to the preparation of novel bio-inspired materials.[6] The hierarchical structure of a gecko toe pad from the macro to the nano level is shown in Figure 4.5. Each toe has a number of layers of tissue called lamellae. These are visible to the human eye and are 400 to 600 μm in width. The lamellae are comprised of foot hairs or setae that are approximately 110 μm in length and 4.2 μm in diameter. Each seta branches into nanoarrays consisting of hundreds

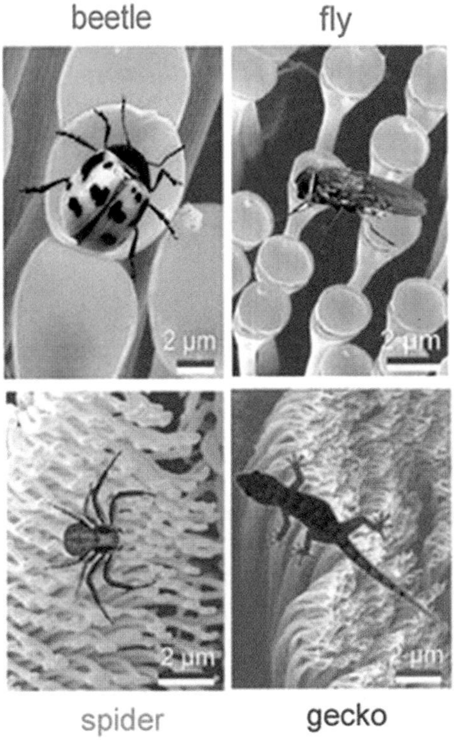
beetle
fly
2 µm
2 µm
2 µm
2 µm
spider
gecko

**Figure 4.4** Micrographs of the fibrillar structures in the hairy attachment pads of beetle, fly, spider and gecko. *Max Planck Gesellschraft*, http://www.mpg.de/bilderBerichteDokumente/multimedial/bilderWissenschaft/2004/05/Gao01e/Web_Zoom.jpeg, 18 November 2009.

of spatulae, which consist of a stalk and a plate-like tip. The stalk is approximately 0.2 µm in length and the tip 0.2 µm in diameter.

The adhesive strength of a single seta has been measured. It was found to be capable of lifting (holding) 20 µN or 20 mg. As the four feet of a gecko have approximately two million setae, they are capable of lifting approximately 41 kg.[7] The adhesive strength of the setae explains the ability of geckos to walk inverted on ceilings. The size and shape of the spatula tip that results in optimum adhesion has been investigated.[8] The critical contact size was found to be ~100 nm (0.1 µm) and below. This size minimizes the effect that surface roughness (irregularity) has on adhesion and results in uniform stress distribution.

#### 4.4.2.2 Wet Adhesion

The toe pad structure of other species has developed to optimize wet adhesion. Wet adhesion differs from dry adhesion in that the toe pad releases a fluid that promotes adhesion. The toe pads of certain amphibians and insect species are

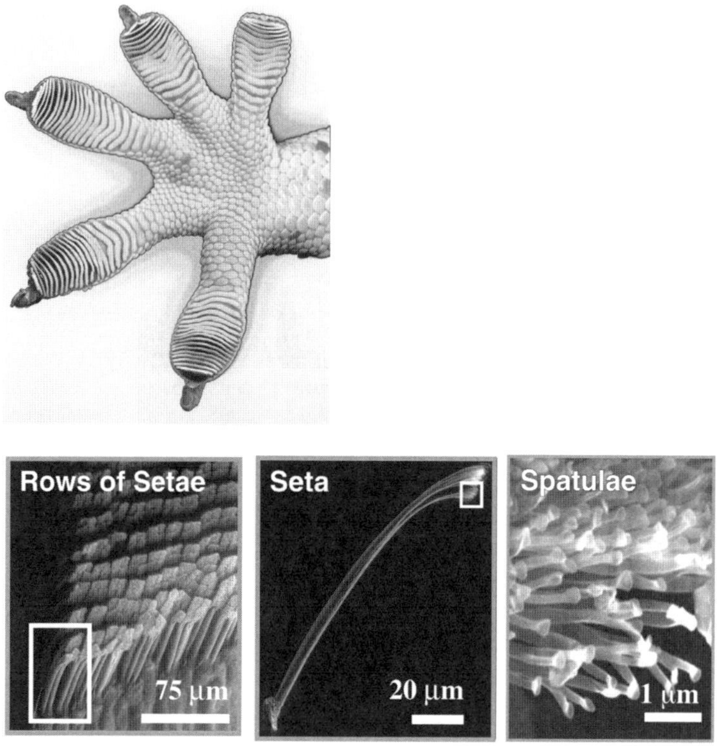
Rows of Setae
75 μm
Seta
20 μm
Spatulae
1 μm

**Figure 4.5** Toe pad structure of a gecko. (Kellar Autumn and Anne M. Peattie, *Integr. Comp. Biol.*, 2002, **42**, 1081–1090, by permission of Oxford University Press.)

examples of wet adhesion structures. A White's tree frog and a low-magnification scanning electron micrograph of a single toe pad, a medium-power micrograph of epithelial cells that constitute the toe pad and a higher power micrograph of the finer (nanoscale) structure on the surface of epithelial cells are shown in Figure 4.6. The micrographs in Figure 4.6 show the nanoscale structure of the epithelial cells at higher magnification. A channel between adjacent epithelial cells and the finer surface structure on the individual epithelial cells is easily seen in the transmission electron micrograph. For comparison, the nanoscale structures on the epithelial cell of a Hylid tree frog are shown in the insert in Figure 4.7. It can be seen that the nanoscale surface structure of the tree frog (family Hylidae) epithelial cells is very similar to that for the White's tree frog.

Scanning electron micrographs of the toe pad epithelium of three families of frogs are shown in Figure 4.8. Comparison of the micrographs indicates the structures are so similar that it would be difficult, if not impossible, to distinguish between the three frogs' families on the basis of their toe pad epithelium. The toe pad epithelium of a cricket is also show in Figure 4.8. The similarity of the toe pad epithelium of the frogs and the cricket is remarkable.[9]

a
b
100μm
c
10μm
d
5μm

**Figure 4.6** (a) Immature White's tree frog. (b) Low-power scanning electron micrograph of the toe pad of a juvenile frog. (c) Medium-power micrograph showing a mucous pore and epithelial cells separated from each other by channels. (d) Higher power micrograph showing finer (nanoscale) structure on the surface of the epithelial cells. (I. Scholz, W. J. P. Barnes, J. M. Smith and W. Baumgartner, *J. Exp. Biol.*, 2009, **212**, 155–162. Reproduced with permission from The Company of Biologists, Ltd.)

The structures associated with dry and wet adhesion are present in a number of species and are examples of convergent evolution. Convergent evolution describes the acquisition of the same biological trait in unrelated lineages. This strongly suggests that the structures associated with these traits have been optimized to perform a function and are excellent candidates for bio-inspired materials or structures.

A bio-inspired synthetic adhesive system has been developed.[6] The adhesive system, based on the attachment pads of beetles, is shown in Figure 4.9. The similarity between the synthetic adhesive and the beetle attachment pad is striking. The bio-inspired adhesive is made using a metal mould. The required surface is embossed as a negative image on a metal surface to prepare the mould. Polymer is poured into the mould, cured and removed.

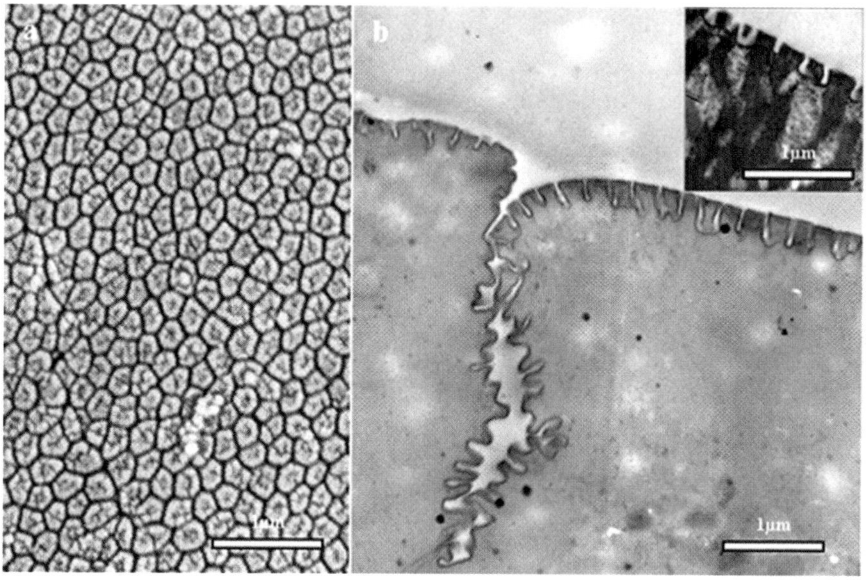
a
b
1µm
1µm
1µm

**Figure 4.7** (a) High-magnification scanning electron micrograph of the surface of toe pad epithelial cells. The length bar is 1 µm. (b) High-power transmission electron micrograph showing a channel separating adjacent epithelial cells and a side view of the nanostructures on the surface of the cell. These are also separated from each other by narrow channels. The inset shows similar nanostructures on a toe pad of the Hylid tree frog, *Scinax ruber*. Note the similarity of the structures at the nanoscale level. (I. Scholz, W. J. P. Barnes, J. M. Smith and W. Baumgartner, *J. Exp. Biol.*, 2009, **212**, 155–162. Reproduced with permission from The Company of Biologists, Ltd.)

## 4.5 Self-healing Materials

Providing that damage is not too significant, animal species have the ability to self-heal. For instance, if one is cut, a complex series of actions that lead to clotting of the blood and ultimately to the healing of the wound is triggered. The sequence of events leading to the clotting of blood in humans is well known and is shown in Figure 4.10. Prothrombin, which is produced in the liver, is modified in a vitamin K dependent reaction. This reaction converts glutamic acids on the prothrombin molecule to gamma hydroxyglutamic acids. Two active sites on the prothrombin molecule are cleaved by an enzyme (Factor X) to produce thrombin. Fibrinogen, a soluble plasma glycoprotein produced in the liver, is converted into fibrin by thrombin. The fibrin is then cross-linked by Factor XIII. This leads to the formation of a clot.

Self-healing polymers are an example of a bio-inspired approach to materials development. Resins containing capsules of uncured resin and a catalyst to promote cure of the resin have been developed.[10] As a crack grows in the

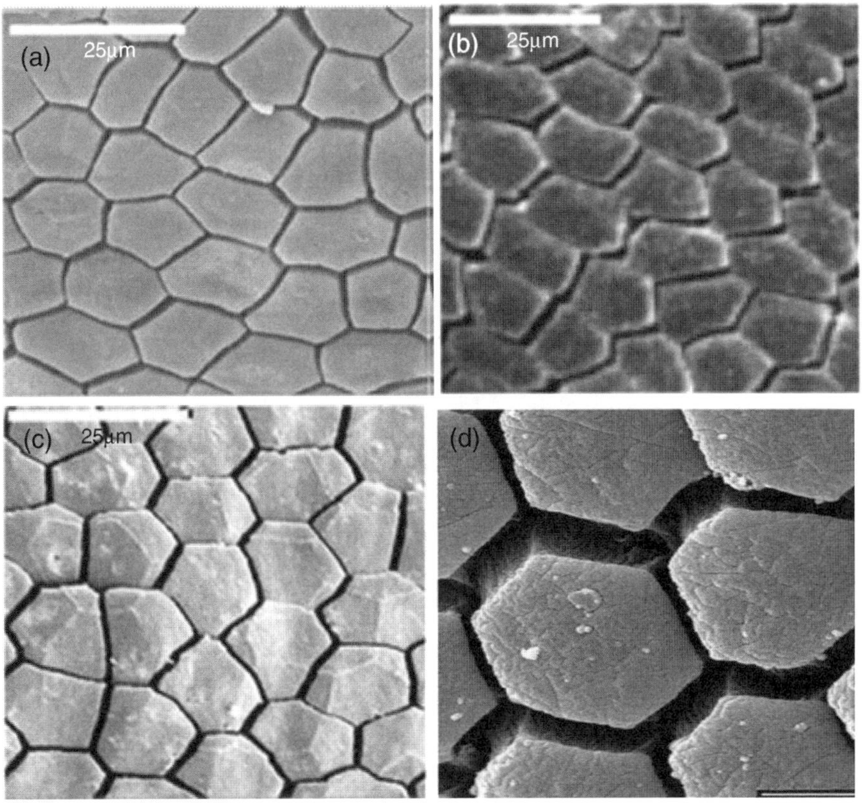
(a) 25μm
(b) 25μm
(c) 25μm
(d)

**Figure 4.8** Toe pad epithelia from three species of frogs (a) *Scinax ruber*, (b) *Rhacophorus schlegelii*, (c) *Mannophryne trinitatis* and a cricket, and (d) *Tettigonia viridissima*. Notice the remarkable similarity in the epithelia. The three frog species evolved separately and therefore the similarity in the toe pad structure is indicative of convergent evolution. Scale bars in a), b) and c) are 25 μm. (Figures (a), (b) and (c) are from Figure 1 in W. J. P. Barnes, *MRS Bulletin*, 2007, **32**(6), 479–485. Reproduced by permission of the MRS Bulletin. Figure (d) is from Figure 3 in S. Gorb, *Proc. R. Soc. Lond. B, Biological Sciences*, **267**(1449), 1239. Reproduced by permission of the Royal Society of Chemistry.)

polymer the capsules of resin are broken. The resin cures when it comes into contact with the transition metal (Grubbs) catalyst incorporated into the polymer. This is shown schematically in Figure 4.11. The load displacement curves for virgin and healed epoxy resin samples are also shown in Figure 4.11. The healed polymer retains a considerable percentage of the load to failure measured for the polymer prior to cracking.

More recently, polymers containing a three-dimensional system of channels, inspired by the circulatory system in living organisms, have been developed. The polymer contains a catalyst and when a crack occurs uncured resin can be moved *via* the channels to the crack site where it cures.[11,12] Toohey *et al.* have

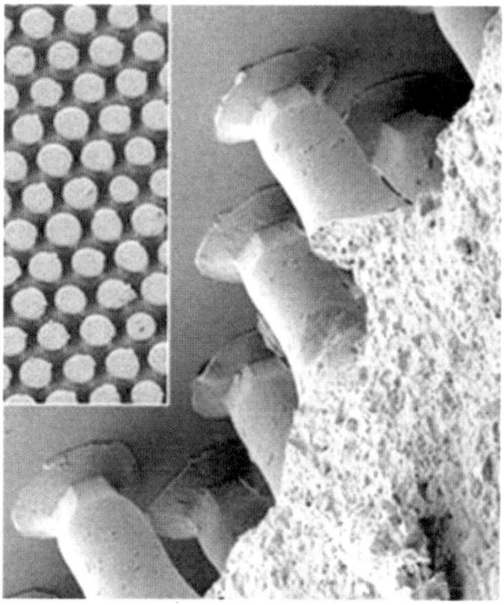

**Figure 4.9** Adhesive system based on beetle attachment pad spatulae. (S. Gorb, M. Varenberg, A. Peressadko and J. Tuma, *J. R. Soc. Interface*, 2007, **4**, 271–275. The Royal Society, London. Reproduced with permission.)

prepared a coating with a three-dimensional network of channels inspired by the microvascular system in animals. The channels contain a resin and the coating a catalyst. When a crack reaches a channel, resin carried in the channels is released and cures when it contacts the catalyst in the resin. This stops the crack.

## 4.6 Superhydrophobic Surfaces

The structure and morphology of plant surfaces responsible for their hydrophobic and self-cleaning properties have been extensively studied. Morphology is defined[13] as 'a branch of biology that deals with the structure and form of an organism at any stage of its life history'. In this text, morphology refers to form and structure and will be addressed from the perspective of how knowledge of morphology can be used to develop bio-inspired materials and structures with desirable properties and functions. The hydrophobic properties result from surface roughness arising from micrometre-sized structures and nanometre-sized wax crystals on these structures. The self-cleaning properties result in particle debris on the surface being carried away as water droplets are shed from the plant surface.[14] A scanning electron micrograph of the surface of a lotus leaf is shown in Figure 4.12. The micrograph reveals that the lotus leaf surface is composed of a pattern of cones approximately 10 μm wide at the base covered with submicron-size wax crystals. The microstructure on the cones plays a significant role in the hydrophobicity of the leaf. When the leaf was

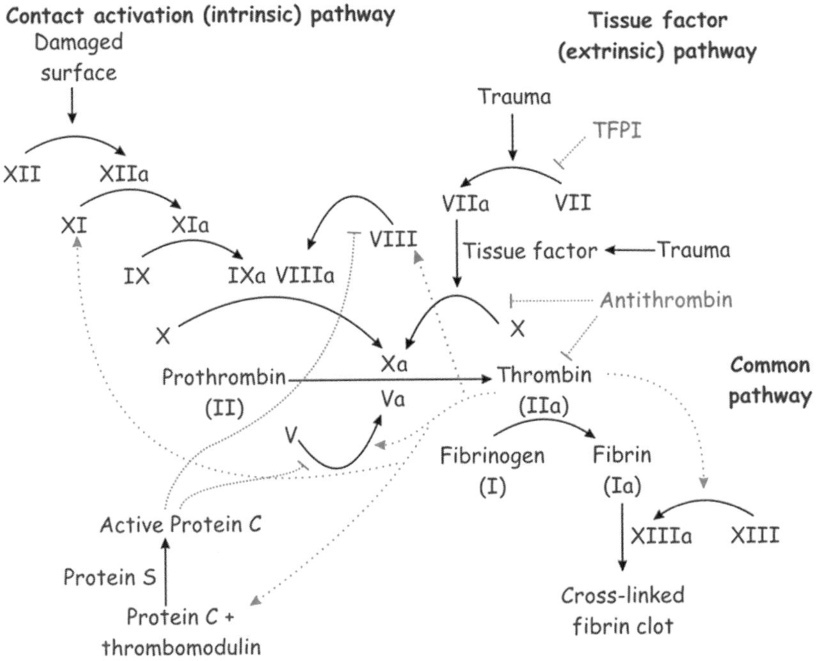

Contact activation (intrinsic) pathway
Damaged surface
XII
XIIa
XI
XIa
IX
IXa VIIIa
VIII
X
Prothrombin (II)
Xa
Va
V
Active Protein C
Protein S
Protein C + thrombomodulin
Tissue factor (extrinsic) pathway
Trauma
TFPI
VIIa
VII
Tissue factor
Trauma
Antithrombin
X
Thrombin (IIa)
Common pathway
Fibrinogen (I)
Fibrin (Ia)
XIIIa
XIII
Cross-linked fibrin clot


**Figure 4.10** Complex pathway leading to the formation of a blood clot. (*Drugs Future*, 2002; **27**(7), 669–683. Copyright © 2002, Prous Science, SA. All rights reserved. Reproduced with permission.)

heated the tiny wax crystals melted (see Figure 4.12), and the hydrophobicity of the surface was decreased.[15,16]

The knowledge of the morphology of plant surfaces and the underlying physical chemistry that results in their hydrophobic properties has been exploited to develop synthetic hydrophobic surfaces. The effect of nano- and micro-patterned structures on the hydrophobicity of synthetic (polymer) surfaces has been reviewed.[17] The structures were based on those found on hydrophobic plant surfaces. The synthetic surfaces with hierarchical structures, that is, surfaces with micrometre-sized asperities covered with nanometre-sized wax crystals, had the highest contact angles and self-cleaning properties similar to those observed for hydrophobic plant surfaces.

## 4.7 Materials

The materials that exist in Nature are as diverse as the structures they constitute. As they are the building blocks for the structures found in Nature, the properties and performance of the structures are tied directly to them.

The bottom-up approach to fabrication results in the self-assembly of the starting materials at the nano- or molecular level and this is carried through to

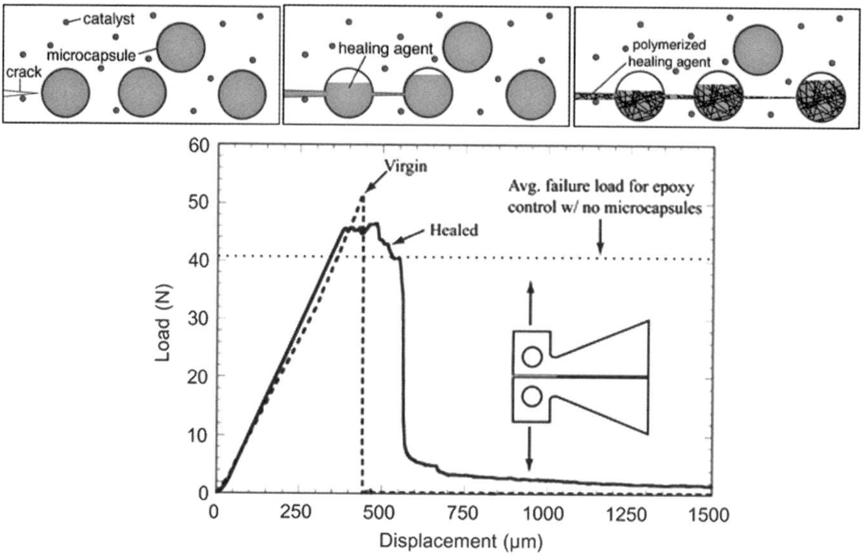
catalyst
microcapsule
crack
healing agent
polymerized healing agent
Virgin
Healed
Avg. failure load for epoxy control w/ no microcapsules
Load (N)
Displacement (μm)
0
10
20
30
40
50
60
0
250
500
750
1000
1250
1500

**Figure 4.11** Top: Graphic representation of how a self-healing resin works. Bottom: Load displacement curves for a self-healing resin prior to and after cracking and self-healing. (S. R. White, N. R. Sottos, J. Moore, P. Geubelle, M. Kessler, E. Brown, S. Suresh and S. Viswanathan, *Nature*, 2001, **409**, 794–797. Copyright 2001. Reprinted by permission from MacMillan Publishers Ltd.)

the final size of the structure of interest. The basic starting materials are simple amino acids, nucleotides, sugars and inorganic compounds such as hydroxyapatite, but the rules used in their formation lead to a vast array of materials with a broad range of properties. The template-based syntheses are derived from the genetic templates found in natural systems. These lead to the formation of polymeric materials, including proteins, nucleic acids and polysaccharides, which have controlled stereochemistry. The resulting materials exhibit a regularity that allows higher order interactions to take place. The higher order interactions are responsible for the self-organization of the constituent materials.

The self-organization of simple molecules into more complex structures is enabled by many weak bonds. These include hydrogen bonding and van der Waals interactions. As was described in the section on hierarchical structure found in bone, the self-assembly of proteins into tropocollagen, then collagen fibres and ultimately collagen fibrils results in spaces where hydroxyapatite can crystallize. Without the incorporation of inorganic crystals, the collagen fibrils would not exhibit the properties associated with bone.

The ability to synthesize materials using a bio-inspired or biomimetics approach has met with limited success. There are a number of reasons for this. Monomeric materials from or used by Nature are stereochemically pure while synthetic monomers are most often racemic mixtures. The stereoisomers in racemic mixtures prevent or inhibit the higher order interactions necessary for

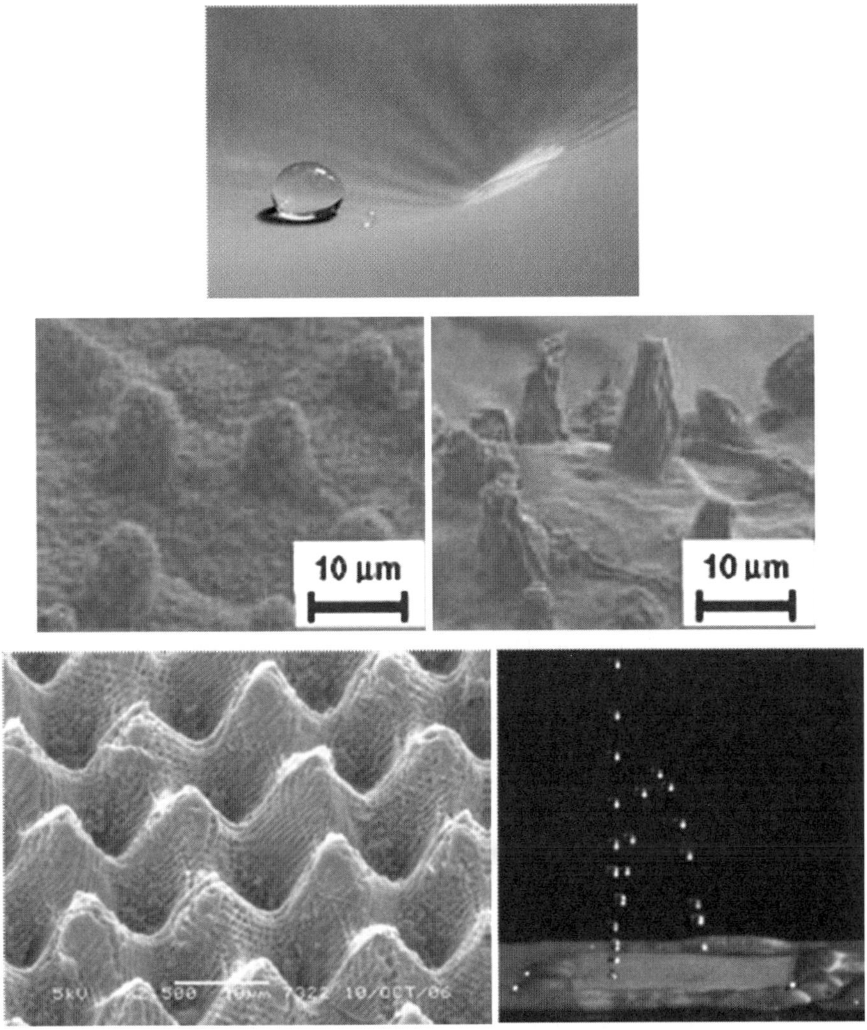
10 µm
10 µm

**Figure 4.12** Top: Water beading on the hydrophobic surface of a lotus leaf. Middle left: Higher magnification SEM of the surface of the lotus leaf showing the microcrystalline wax on the cone structures. Middle right: Cone structure after heating; the microcrystalline wax structure has been destroyed. Bottom left: SEM of bio-inspired cone and submicron fine structure on a hydrophobic polypropylene surface. Bottom right: Water droplets are repelled from the hydrophobic polypropylene surface. (Top photo and bottom micrographs: M. Groenendijk, *Laser Technik Journal*, May 2008, 44–47. Copyright Wiley-VCH Verlag GmbH & Co. KGaA. Reproduced with permission. Middle micrographs: Y. T. Cheng, D. E. Rodak, C. A. Wong and C. A. Hayden, *Nanotechnology*, 2006, **17**, 1359–1362. IOP Publishing Limited. Reproduced with permission.)

self-assembly into more complex molecules and materials. The polymers arising from natural and synthetic sources are also different. Natural polymers exist in a broad range of forms from mono- to highly diverse block polymers while synthetic polymers are usually mono- or diblock polymers.

The rate of polymerization also varies significantly. A slow, in a comparative sense, template-driven approach to material fabrication is used by Nature whereas laboratory syntheses are comparatively fast. The processing and assembly conditions used by Nature are also different from those used for the majority of synthetic materials. Natural synthesis (materials formation) takes place in water and the resulting polymers assemble through hydrogen bonding and other interactions such as van der Waals forces into more complex structures. Synthetic polymers are most often prepared in organic solvents and in most cases they do not self-assemble. The interactions between individual polymer chains are generally the result of entanglements and not hydrogen bonding and van der Waals forces.

Nature uses materials that are available in the environment. Plants use water, carbon dioxide, chlorophyll and sunlight to produce the compounds required for growth. Animals use complex carbohydrates, proteins, inorganic salts (minerals) and water to produce the myriad of compounds and structures required for growth and continuing life. Although the ability to live and thrive in an environment using very basic starting materials is a very attractive concept, natural materials and systems do have a number of limitations. Many of these are related to the environment in which they can exist and thrive. Often plants and animals species can exist in a very narrow temperature range, are susceptible to high and low pH, have limited resistance to a number of chemicals and require water for their continued existence.

These limitations make the syntheses of bio-inspired materials with compounds (chemical species) which are not subject to similar limitations very attractive. This assumes that the relationship between the structure and properties (or performance) of a natural system or subcomponent of that system is known, the performance of the system is not material specific and that the structure can be made from materials that have improved resistance to the environment where it will be used. Super hydrophobic surfaces were discussed earlier in this chapter. The parameters that result in their performance have been illuminated. However, the materials would not perform in certain environments, for instance, where they might be exposed to organic solvents or be subject to abrasion from sand and dust particles. If the structure responsible for the inability of water to wet the surface can be produced on a plastic or metal surface, then the super hydrophobic and self-cleaning properties will exist on a more robust substrate.

## 4.8 Conclusion

As was noted at the start of this chapter, structure is the basis of the wide range of properties found in Nature. This is true at the component level, the system level and the system of systems level.

The examples discussed in this chapter were chosen to illustrate several points. The first is how relatively simple starting materials such as collagen and hydroxyapatite can result in a material (bone) with properties unlike those of the starting materials. Structure is maintained from the atomic and molecular level through the micro level to the final scale of the material or system.

The second is the challenge in mimicking the bottom-up approach used by Nature to produce materials and systems. To date, the use of the top-down approach to produce bio-inspired materials and systems has found more success than the bottom-up approach. The successes arising from the top-down approach have resulted from an in-depth understanding of the relationship between the observed properties of a natural material or system and its structure.

The third is the multifunctional nature of natural materials. The surface of a natural system may provide protection from the environment, stealth, be involved in collection and sampling and be an integral part of the system responsible for mobility. In the context of autonomous sensing systems, the ability to create bio-inspired multifunctional materials will allow sensor developers to address several of these concerns with a single material. The multifunctional or integrated nature of materials and systems may also impact (reduce) power requirements.

Natural materials and systems are not without limitations and there are opportunities, if not requirements, to address these limitations through the use of more robust materials.

In natural sensing systems, there is a continuum that includes sampling and collection, specific and sensitive reception and transduction of analytes. In general the components of the system responsible for these functions are integrated and there is no easy way to separate the components or their function. In the next three chapters; Chapter 5 – Collection and Sampling, Chapter 6 – Receptors and Surfaces and Chapter 7 – Sensing and Transduction, these functions will be discussed with this in mind.

# References

1. M. J. Buehler, *Nanotechnology*, 2007, **18**, 295102.
2. S. Weiner and W. Traub, *The FASEB Journal*, 1992, **6**, 879.
3. J. M. Benyus, *Biomimicry – Innovation Inspired by Nature*, Perennial/An Imprint of Harper Collins Publishers Inc., New York, 2002.
4. *MRS Bulletin*, 2007, June, 32.
5. K. Autumn, M. Sitti, Y. A. Liang, A. M. Peattie, W. R. Hansen, S. Sponberg, T. W. Kenny, R. Fearing, J. N. Israelachvili and R. J. Full, *Proc. Natl. Acad. Sci. USA*, 2002, **99**, 12252.
6. S. Gorb, M. Varenberg, A. Peressadko and J. Tuma, *J. R. Soc. Interface*, 2007, **4**, 271.
7. K. Autumn, Y. A. Liang, T. S. Hsieh, W. Zesch, W. P. Chan, T. W. Kenny, R. Fearing and R. J. Full, *Nature*, 2000, **405**, 681.

8. H. Gao and H. H. Yao, *Proc. Natl. Acad. Sci. USA*, 2004, **101**, 7851.
9. W. J. P. Barnes, *MRS Bulletin*, 2007, **32**, 479.
10. S. R. White, N. R. Sottos, J. Moore, P. Geubelle, M. Kessler, E. Brown, S. Suresh and S. Viswanathan, *Nature*, 2001, **409**, 794.
11. A. Nagar, E. Anderson, P. Rao and S. Narang, in *SPIE 14th International Symposium on Smart Structures and Materials & Nondestructive Evaluation and Health Monitoring*, San Diego, California, 2007, paper 6525-46.
12. K. S. Toohey, N. R. Sottos, J. A. Lewis, J. S. Moore and S. R. White, *Nat. Mater.*, 2007, **6**, 581.
13. *McGraw-Hill Dictionary of Scientific and Technical Terms*, McGraw-Hill, Toronto, Ontario, Canada, 6th edn, 2002.
14. C. Neinhuis and W. Barthlott, *Ann. Bot.*, 1997, **79**, 667.
15. M. Nosonovsky and B. Bharat, *Microelectron. Eng.*, 2007, **84**, 382.
16. N. J. Shirtcliffe, G. McHale, M. I. Newton, G. Chabrol and C. C. Perry, *Adv. Maters.*, 2004, **16**, 1929.
17. B. Bhushan, Y. C. Jung and K. Koch, *Phil. Trans. R. Soc. A*, 2009, **367**, 1631.

CHAPTER 5

# Collection and Sampling

## 5.1 Introduction

A prerequisite for the detection of a stimulus (analyte), whether it be optical, aural, chemical, biological, thermal or magnetic, is that it be present in sufficient concentration or have sufficient intensity to allow it to be detected and differentiated from other interfering stimuli. This requires that the analyte be collected or sampled from its environment. This is often a very difficult problem for two reasons. The first is that the concentration of the compound or intensity of the stimulus may be very low and the second is that the compound or stimulus of interest may be a component of a complicated mixture or be present among a number of other stimuli. To overcome these problems Nature has developed collection, sampling and ultimately detection (receptor and transducer) systems that are sensitive and specific.

Figure 5.1 illustrates the range of media from which a compound may have to be sampled and collected. It also delineates some of the challenges involved in developing sampling and collection systems for use in sensing systems. These challenges range from stealth concerns, for instance, is the device easily seen or heard, to size, power requirements, environmental concerns and how to move the analyte to the transducer or detector. There is a wealth of information that can be gained from natural systems that can be used to address these challenges.

Natural systems exist that can sample efficiently from air, water and solids. In all three media, the compound of interest can be present in very low concentrations and as a component of a complex mixture. Natural systems have evolved to maximize both sensitivity and specificity for the analyte of interest. Each of the media in turn presents unique challenges from the perspective of

Bio-inspired Materials and Sensing Systems
By Peter Biggins, John Hiltz and Anne Kusterbeck

Published by the Royal Society of Chemistry, www.rsc.org

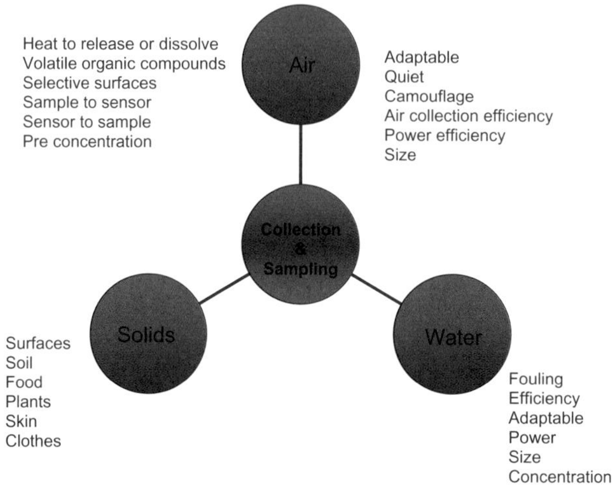
Heat to release or dissolve
Volatile organic compounds
Selective surfaces
Sample to sensor
Sensor to sample
Pre concentration
Air
Adaptable
Quiet
Camouflage
Air collection efficiency
Power efficiency
Size
Collection & Sampling
Solids
Surfaces
Soil
Food
Plants
Skin
Clothes
Water
Fouling
Efficiency
Adaptable
Power
Size
Concentration

**Figure 5.1** Schematic illustrating the factors affecting collection and sampling from air, water and solids.

sampling and collection. Airborne samples may be present in extremely low concentrations and require a concentration step prior to detection or utilization. Fouling of the surface or structure responsible for collection and sampling may compromise the ability of a sensor system to detect the analyte at the required sensitivity. The interaction of a compound with a solid sample may require the use of heat or chemicals to release it prior to sampling. Collection and sampling in natural systems is also carried out in an energy-efficient manner. The surfaces or structures responsible for collection and sampling often have a very short refresh time, that is, the time between collection and detection cycles. All these factors make knowledge of how natural systems collect and sample from the environment of great interest to those who want to develop improved sampling and collection devices. These are an integral part of the autonomous sensing systems discussed in Chapter 3.

In this chapter, collection and sampling of stimuli (analytes) will be discussed.

## 5.2 Approaches to Collection and Sampling

Scientists have developed sampling and collection techniques for substances from gaseous (air), liquid (water and other fluids) and solid samples. The technique used depends upon the concentration of the compound in the sample and the detection limit of the compound. If the concentration of an airborne substance is low, large volumes of air may have to be processed to ensure that enough is collected to allow detection. The processing often involves moving air through a filter if the substance is a solid or through a chemical sorbent (extraction media) if it is in the vapour state. Activated charcoal, molecular sieves and chemical sorbents such as Tenax® and Chromasorb® are used to extract the species of interest from the matrix that contains it. The compound must then be removed from the sorbent (extraction phase) prior to subsequent preparation and detection. This may require the use of a solvent or heat to remove it from the sorbent. If the extraction medium is not specific for a particular compound then a separation step may be required prior to analysis. Similarly, the collection of a compound from liquid and solid samples may require the processing of large volumes (or masses) of material and subsequent workup prior to detection.

Static and dynamic headspace analyses are used to sample volatiles from air, liquid and solid samples. In static headspace analysis, the air in the vicinity of the sample is taken. This is similar to one sniffing to detect the odour of a substance. In the dynamic headspace analysis, the air in the vicinity of a sample is moved through a chemical trap or a cold trap over a period of time to concentrate the analyte of interest. This sampling technique then requires that the sample be removed from the trap, generally by heating, prior to analysis. More recently, membrane extraction techniques such as solid phase micro-extraction have been developed to facilitate sample collection. In this technique, a polymeric membrane is coated on a hollow fibre. This is exposed to the headspace above a sample or directly into the sample and analytes are absorbed by the coating. The absorbed molecules can then be removed by heating. No solvents, chemical traps or cold traps are required.

The techniques described above have been developed to collect sufficient sample to allow detection or to speed the sampling and collection of compounds and reduce the time and cost associated with the sampling and collection phase of the detection process. From the defence perspective, both speed and cost of detection are critical especially for systems in operational areas. In the next section natural collection and sampling systems and examples of systems inspired by these natural systems will be discussed.

### 5.2.1 Collection and Sampling Tools

A number of the collection and sampling tools that we use in laboratories today might be described as being bio-inspired or biomimetic. Even if the inspiration for these tools was not a natural system, the similarity between them and those found in natural systems is remarkable. In the following paragraphs sampling tools found in Nature are compared to man-made tools that have similar functions. Many of the natural tools are found in the insect world and are mouth parts that have been specialized for processing food from a wide variety of sources.

The syringe is widely used in scientific and medical applications and has a parallel in the insect world. The syringe can be used to both inject and extract fluids. The Assassin bug (family Reduviidae, suborder Heteroptera, order Hemipteras) and other bugs in this genus inject a paralysing digestive fluid into their prey *via* a syringe-like mouth part. They then use the same part to suck up the digested contents of the prey.

The Desert Locust (*Schistocerca gregaria*) has very sharp cutting appendages which are used to attack food in a cut-snip mode; the same way we use scissors. Predatory dragon flies Odonata have specialized jaws to grab prey in the same way we would use forceps to clamp on to an object. Lobsters (*Homarus americanus*) have developed powerful appendages that are used to crack open hard mollusc shells. The praying mantis, *Mantodea mantidae*, has strong limbs for gripping prey in the same way we would use pliers. The horseshoe crab (*Limulus polyphenus*) uses the rasping action of its claws to tear up food prior to moving the food to its mouth.

The common bluebottle fly (*Calliphora vicina*) vomits digestive fluids over its food source. It then uses its sponge-like mouthparts to suck up the digested food, in the same way we would use a sponge to soak up fluid. Moths and butterflies sip nectar from plants using a straw-like feeding tube or proboscis in the same way a scientist would draw a sample using a pipette.

Organisms have also developed the ability to sieve material from liquid (water). The legs of prawns (*e.g.* Aytid prawns) have fine hair structures that sieve food particles from the water. Barnacles move their limbs, containing fine hairs, in a sweeping motion to capture food particles from the water. In an analogous manner, scientists use sieves of various mesh sizes and fibrous mats to sample solids from air and water samples.

## 5.3 Natural Sampling System – Olfaction

Olfaction is an example of a natural sensing system. The olfactory system has evolved in a way that enables the collection, detection and differentiation of molecules (odours) from the environment. The upper portion of the nasal passages of mammals has a mucous lining. Molecules passing through the nasal passages dissolve in this lining. They then diffuse through the lining to olfactory receptors where they bind to proteins and are detected. In insects the molecules penetrate into cuticle pores of the chemosensory sensilla located on various body parts including their antenna and tarsa. The odour-binding proteins are located there.

The ability of olfactory systems to detect and differentiate odours varies considerably in Nature. Dogs have a more sensitive sense of smell than humans. There are two reasons for this. The surface area dedicated to olfaction, the olfactory epithelium, is larger and the density of innervated sites in the olfactory epithelium is higher in dogs than in humans. On average humans have approximately 10 $cm^2$ of olfactory epithelium whereas some dogs have 170 $cm^2$ of olfactory epithelium. Dogs can also have 100 times more receptors (olfactory neurons) per $cm^2$ of olfactory epithelium than humans. The sensitivity of a dog's nose is compared to explosive detection equipment in Figure 5.2.

Specificity related to smell arises from the interaction of odour molecules with specific receptor proteins. Each protein recognizes only a particular aspect of an odour molecule or class of odour molecules. Mammals have about 1000 functional genes expressing for odour reception. Only a portion of these are functional odour receptors and humans have fewer than other mammals. In addition, each olfactory receptor neuron has only one functional odour receptor. If the odour molecule can interact with the receptor, the nerve cell will respond. Several theories, including the weak-shape or odotope theory and the vibration theory[1,2] have been proposed to explain how odour coding and

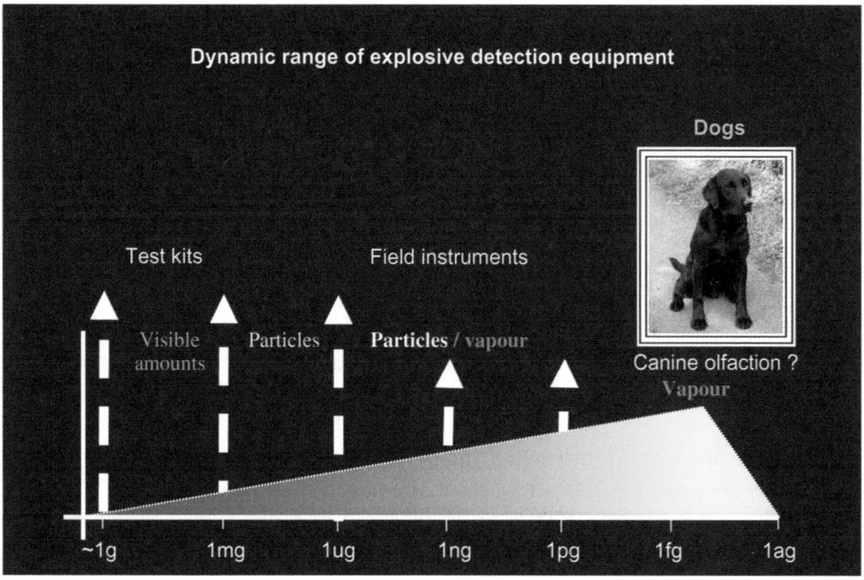

Dynamic range of explosive detection equipment
Dogs
Test kits
Field instruments
Visible amounts
Particles
Particles / vapour
Canine olfaction ?
Vapour
~1g
1mg
1ug
1ng
1pg
1fg
1ag


**Figure 5.2** Comparision of a dog's nose with conventional detection systems for vapour detection (courtesy of A. Hutchinson).

perception takes place. The odotope theory proposes that different receptor neurons detect the constituent parts of odour molecules and the inputs from a number of receptor neurons are combined to form an olfactory perception. The vibration theory proposes that an odour receptor detects the infra-red frequencies of an odour molecule that has interacted with it *via* electron tunnelling.

## 5.4 Bio-inspired Sampling System – Electronic Nose

There has been considerable research directed at the development of artificial or electronic noses. A recent US Defense Advanced Research Projects Agency (DARPA) project titled RealNose[3] has the objective of building a mechanical nose that closely simulates the entire canine olfactory system. This system is not only bio-inspired but parts of the proposed nose are biomimetic. The project incorporates earlier work carried out at Penn State University on understanding the flow of odorants through the nasal cavity and over the olfactory epithelial cells in dogs. Computational fluid dynamics modelling work, done as part of the Penn State study, gave insight into how odours are transported to the olfactory region and indicated that the flow is very smooth. This smooth flow provides a consistent transport of odorant molecules to the olfactory receptors, maximizes their flow to the receptors and produces a deposition pattern that is interpreted by the animal's brain.[3] One approach proposed in the RealNose project for the collection of the odour molecules is to attach receptors to carbon nanomaterials to mimic receptor proteins found in the olfactory epithelium of dogs and other mammals.

Although the DARPA-funded research described in the preceding paragraph is promising, especially with respect to ensuring that odour molecules reach the sensing surfaces, attempts to build artificial noses capable of detecting a broad range of molecules have met with limited success. A recent review notes that the original approach to improving electronic noses was to increase the number of individual detectors. However, it has become apparent that it is not only the number of odour receptors but also their selectivity and sensitivity to individual molecules that are critical to olfaction.[4] Rock *et al.* suggest that addressing these concerns can be done in several ways. One is to develop new sensing materials, such as DNA-based, molecularly imprinted polymers and immobilized natural receptor materials, to increase both sensitivity and selectivity for odour molecules. The other is to pretreat (preconcentrate) the sample to increase sensitivity or use filters or separation techniques to improve selectivity.

Another approach to improving the performance of artificial noses, based on mimicking the mucous layer in the nose, has been investigated.[5,6] In the human nose, scent (odour) molecules dissolve in the mucus layer and based on their chemistry arrive at the odour receptors at different times. This time (temporal) information, in conjunction with the magnitude of the response (spatial) for scent molecules, is used to differentiate between smells. The research team applied a 10-micron-thick layer of polymer on a planar chromatographic channel to provide some separation (retentive delay) of scent molecules. Forty sensors were placed along the channel. Each of the sensors had one of ten different sensing materials. The device provides both spatial and temporal information on scent molecules. The spatial, temporal and spatial-temporal data for both simple (toluene, ethanol) and complex (milk, banana, peppermint essence) scents were analysed using principal components analysis. The spatial-temporal data gave better results than either the temporal or spatial data alone.

## 5.5 Bio-inspired Materials for Collection and Sampling

In the discussion of olfaction and research directed at developing artificial or electronic noses, the development of bio-inspired materials, such as molecularly imprinted polymers (MIPs), with enhanced selectivity and sensitivity to analytes was mentioned. MIPs and several other bio-inspired materials used in sampling and collection will be discussed in this section.

### 5.5.1 Molecularly Imprinted Polymers

Molecular imprinting is a bio-inspired approach to preparing synthetic polymers with selective recognition properties for particular analytes. The goal of molecular imprinting is to prepare polymer-based substrates with selectivities similar to those found for naturally occurring systems.[7]

The classic approach to the preparation of molecularly imprinted polymers involves the copolymerization of functional and cross-linking monomers in the presence of a target molecule (the imprint molecule). The polymerization is often carried out in a solvent (porogen). The solvent acts not only to dissolve the monomers and imprint molecules but also mediates interactions between the functional groups and the imprint molecule, determines the timing of phase separation during the polymerization which affects morphology and generates a porous structure in the MIP.[8] The porous structure facilitates the extraction of imprint molecules and subsequent ingress of the molecule of interest to the imprinted sites.

The imprinting of the polymer depends on the interaction of the target molecule with the functional monomer. This interaction can take place in one of two ways. In the first, a complex between the imprint molecule and the functional monomer is formed based on weak intermolecular interactions (ionic or hydrophobic interactions, hydrogen bonding and metal coordinations) prior to polymerization. This approach is shown schematically in Figure 5.3. Mosbach

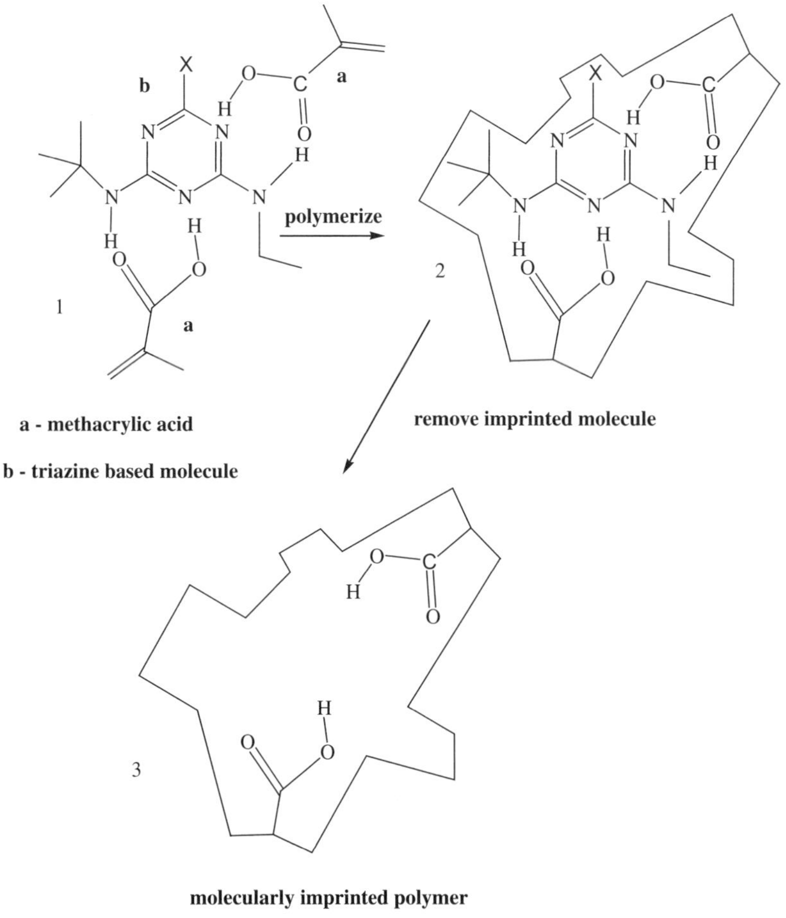
polymerize
1
2
3
a - methacrylic acid
b - triazine based molecule
remove imprinted molecule
molecularly imprinted polymer

**Figure 5.3** Preparation of a molecularly imprinted polymer.

and co-workers pioneered the development of this approach.[9] Following polymerization, the MIP is washed with solvent to remove the imprint molecule. In the second, covalent bonds are formed between the functional monomer and the imprinted molecule to yield a polymerizable derivative of the imprinted molecule. This derivative is then polymerized. Wulff and colleagues pioneered the development of this approach.[10] When using this synthetic approach, covalent bonds must be broken prior to removal of the imprint molecule.

Molecularly imprinted polymers with excellent affinity and specificity for a large range of compounds have been synthesized. These include small organic molecules, pharmaceuticals, pesticides, amino acids and peptides, nucleotide bases, steroids and sugars. Compounds with larger structures present a more difficult challenge.

Although the collection and sampling aspects of MIPs show great promise for sensors or sensing systems, there are several critical issues to the design of MIP-based sensors.[11] These include the development of sensitive transducers capable of monitoring the binding process and transforming it into a signal, the development of MIPs capable of interacting with the target compound under the desired conditions and with the required affinity and specificity and the integration of the MIP with the transducer.

The response time of sensors is an important parameter. If the compound has to diffuse into the MIP, this can limit the physical form of the MIP, for instance, the thickness of a film or the size of a powder that can be used in the sampling device. The physical form of the MIP will also influence the reversibility of the binding process and ultimately the reusability of the sensor. Questions such as how long will it take to remove the analyte? and how complete will the removal process be? will have to be considered.

### 5.5.2 High Surface Area, Highly Porous Materials

The aim of porous materials research is to control the size, shape and uniformity of the pores and the atoms and molecules that define them[12] and prepare materials that mimic the naturally occurring porous materials found in Nature. These naturally occurring porous materials have hierarchical structures with order on the nanometre (atom and molecule) level through the micrometre and onto the millimetre level.

Zeolites are an example of natural porous minerals. Zeolites are hydrated alumino-silicate minerals with an open, porous structure and are members of the molecular sieve family. Because of their molecular-sized and regular pore structures, these minerals can separate molecules on the basis of size. Some typical zeolite structures are shown in Figure 5.4.[13,14]

Galen D. Stucky at the University of California, Santa Barbara, defines high surface area porous materials technology as follows:

'High surface area materials (300–2000 $m^2 g^{-1}$) technology is directed at creation of inexpensive low bulk volume/area media for applications that require rapid and responsive sampling, selective separations, catalytic processing, enhanced chemical reactivity, or 3-D packaging of supported or entrained

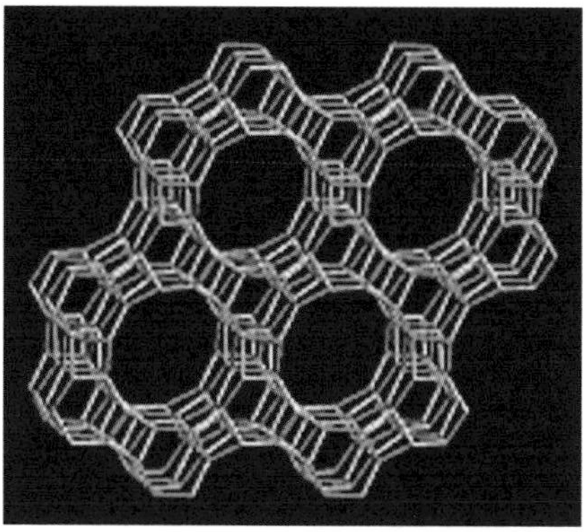

Zeolite AFI- Tetrapropylammonium Fluoride AlPO-5

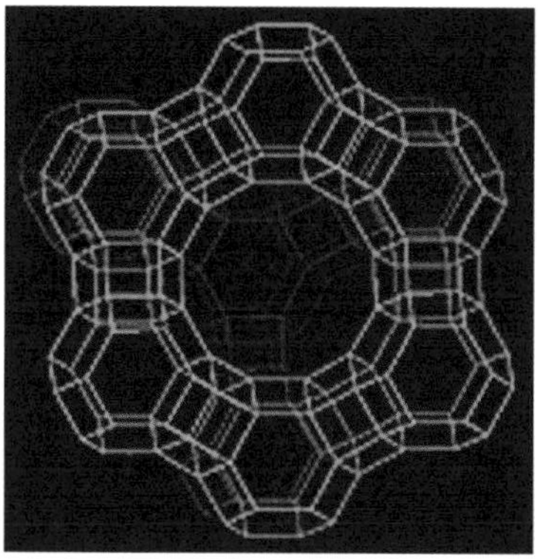

Zeolite FAU - Faujasite

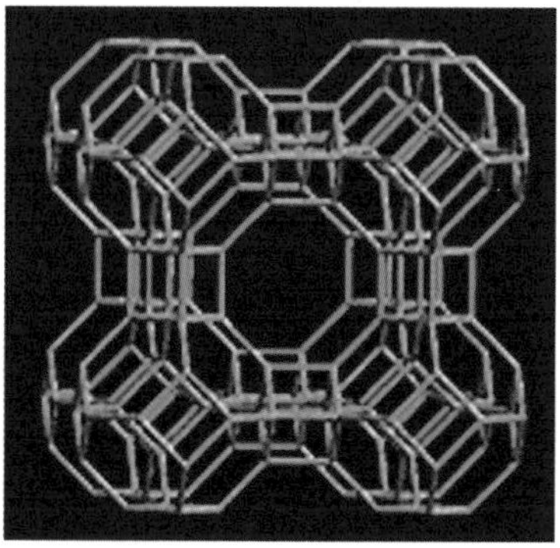

Zeolite LTA - Linde Type A

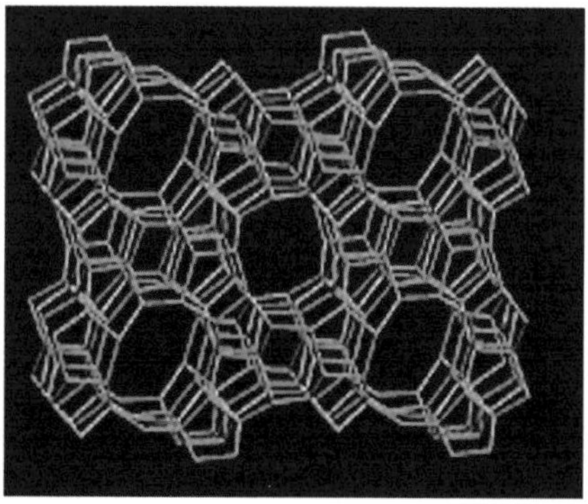

Zeolite MFI - Tetrapropylammonium ZSM-5

**Figure 5.4** Some typical zeolite structures.[13]

nanoscopic structured species. The high surface area of these materials provides a means to achieve detection sensitivities in the ppb range or to rapidly sample and chemically process large volumes of reactants.'

Fractional void space in porous materials is in the range of 0.9 with pore volumes ranging from $0.6\,cm^3\,g^{-1}$ to $2.7\,cm^3\,g^{-1}$. The ability to create three-dimensional patterning and periodicity results in the optimum surface area/volume control, useful access space and control of structure and properties at the nanoscale level. The ability to control pore size at the Angstrom level ($10^{-10}$ m) results in materials with selectivities for separation or catalytic processes that are several orders of magnitude better than previously possible.

One of the potential applications for this technology is in chemical sensing using films, spheres or fibres.[15,16] There are a number of control parameters required for this type of application and they are similar to those of enzymes in biocatalysts. Key variables include: molecular recognition parameters, transition state lifetimes, sorption and desorption rates, the ability to functionalize the surface, chemical and mechanical stability of materials, defined defect structure and interface chemistry from both a synthesis and a composite property perspective.

A form of molecular imprinting defines the topology of high surface area materials at the nanostructure level. The imprinting can involve single molecules or organized arrays of molecules. The actual forms of the materials created depend on several factors. These include the relative kinetics and thermodynamics of the polymerization of the molecules that make up the surface, the interface interactions between these molecules and those responsible for the creation of the high surface area and surface nanostructure and the intramolecular interactions of these molecules.

The synthesis of high surface area macroscale structures (fibres, films, beads and millimetre-sized single crystals) with monodisperse porosities has met limited success. Processing of high surface area materials into forms or shapes (films, fibres, spheres or macroscale patterned structures) that can be used for a particular application is a challenge. A goal in this area is to design structures based on nanoscale molecular assembly that incorporate control of structure and properties on a space and time basis.

Progress in generating monodisperse, high surface area materials containing 3-D periodic arrays of pores and cages with dimensions greater than $10^{-9}$ m has been slow. Attempts to prepare materials that bridge the nanoscale ($<10^{-7}$ m) to mesoscale (up to $10^{-3}$ m to $10^{-2}$ m) dimensions have resulted in loss of short-range order at the nanoscale level.

There are a number of other parameters that must be addressed to enable the use of high surface area materials. These include the mechanical and chemical stability of high surface area, large pore size materials, functionalization, defect chemistry, diffusion processes and the determination of the pore and cage structures in these compounds.

High surface area or highly porous materials show tremendous promise for use in collection and sampling applications. The control of pore size and functionalization of the surface to promote or enhance selectivity will produce materials with a broad range of uses as sampling and collection components of sensing systems.

### 5.5.3 Polysilsesquioxanes

Bridged polysilsesquioxanes are hybrid organic-inorganic materials prepared by sol-gel processing of monomers containing an organic bridging group and two or more trifunctional trihalosilyl or trialkoxysilyl groups. The organic bridging portion of the monomers can vary significantly and includes rigid arylene and acetylene groups, flexible alkylene groups containing up to 14 methylene groups and functionalized amines, ethers, sulfides and carbonates. Hydrolysis of the monomers and subsequent condensation steps lead to the formation of polysilsesquioxanes. By varying the nature of the organic bridging group, the bulk properties of the resulting polymer, such as porosity, thermal stability, chemical resistance and hydrophobicity, can be changed.

The organic bridging groups affect the size, shape, geometry and functionality of the polysilsesquioxane monomers and therefore bulk properties, such as porosity, of the polymer.[17] In amorphous bridged polysilsesquioxanes, the compliance of the network is an important contributing factor to porosity. The compliance depends on the degree of condensation of the silicon and the flexibility of the bridging group. Sol-gels prepared under acidic conditions tend to have lower degrees of condensation than those with basic catalysts. For instance, polysilsesquioxanes prepared from monomers with flexible bridging groups, such as long chain alkylenes, under acidic conditions are susceptible to collapse and can result in non-porous xerogels or thin films.

Less compliant networks retain their porosity on drying. Materials with less compliant networks are prepared using basic catalysts and from monomers with less flexible bridging groups. These materials can have surface areas up to $1800\,m^2\,g^{-1}$. Micropores (widths less than 2 nm) make a significant contribution to the surface area. Mesoporous xerogels (pore widths from 2 nm to 50 nm) can be prepared using a basic catalyst and alkylene bridging groups containing up to 10 carbons.

The organic bridging group can also be used as a template to create porosity. The organic bridging group occupies space in the polysilsesquioxane until burned away, oxidized, hydrolysed or changed by chemical rearrangements. This process leaves pores that are roughly of the same size and shape as the bridging group. Thermolysis of non-porous xerogels has produced porous silica gels, while the oxidation of porous xerogels resulted in gels with larger pore sizes. This technique has been used to template porosity in silica membranes.[18] Chemical reactions can also be used to modify, cleave or remove portions of bridging groups in polysilsesquioxanes. For instance, hydrolysis of carbamate linkages in the bridging group has been used to prepare aminoalkyl functionalized polysilsesquioxanes.[19]

The high surface area and chemical functionality of bridged polysilsesquioxanes makes them candidate materials for separation media, as pore templates and structural materials for membrane materials, as catalyst supports and as metal and organic absorbents.

### 5.5.4 Dendrimers

Dendritic architecture is found at all size scales in Nature; at the metre scale in tree branches and roots, at the centimetre and millimetre level in structures such as the lungs, kidney and liver, at the micrometre scale in cerebral neurons and at the molecular level in compounds such as glycogen, amylpectin and proteoglycans.[20]

Dendritic macromolecules are synthetic compounds with highly branched three-dimensional structures that mimic those found in highly branched natural molecules and systems. Once these compounds reach a certain size they adopt a globular shape that is the direct result of their branching pattern. Dendritic macromolecules can be separated into three broad types; dendrimers, hyperbranched polymers and hybrids of dendrimers and hyperbranched polymers. Dendrimers differ from hyperbranched polymers and dendrimer-hyperbranched polymer hybrids in that they have highly regular structures and are usually monodisperse. This is unusual for synthetic macromolecules.

Dendrimers can be made in one of two ways. The first uses a divergent synthesis and the second a convergent synthesis. The divergent synthesis starts with a single, multifunctional core element. The building blocks (or branch cells) that react with the core unit each have a branch junction. Further reaction of building blocks with functional groups on the branch cells leads to the formation of the next generation of branching. As each successive generation is added, the reaction slows due to increased crowding at the molecule's periphery. In a convergent synthesis, the branches of the dendrimer are prepared and then reacted with the core element. Steric hindrance limits the number of generations that can be incorporated into dendrimers prepared using a convergent synthetic strategy.

The surface of a dendritic polymer has a number of terminal or surface groups. The number of these depends on the functionality of the core molecule

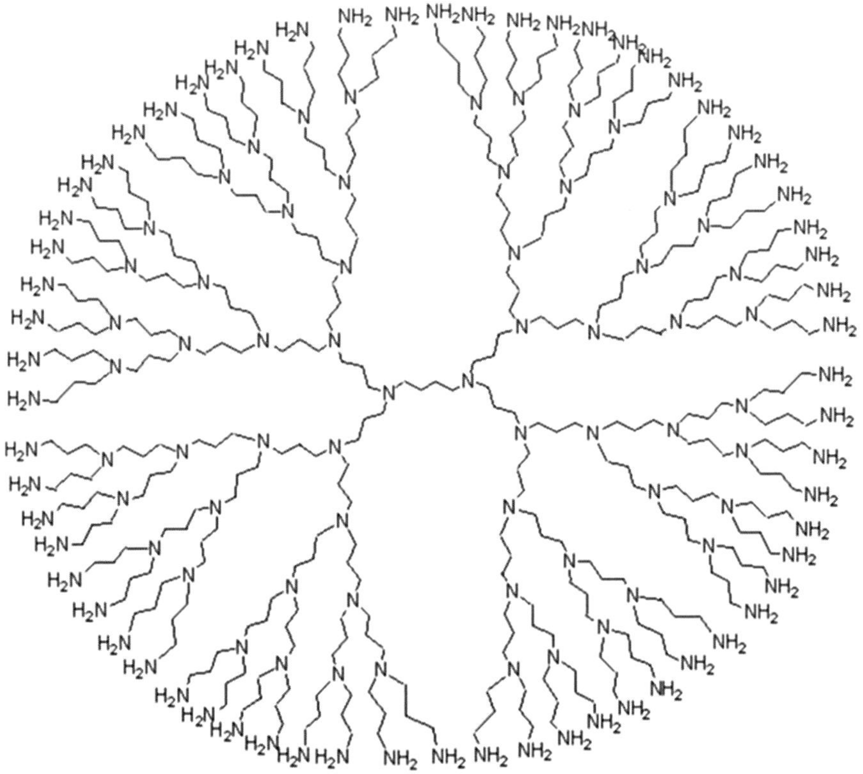

**Figure 5.5** A fifth-generation dendritic polymer.

and the building blocks and can be calculated for each generation of the dendrimer. For example, if the core molecule is trifunctional and the building blocks are bifunctional, the number of terminal groups for the fourth generation of a dendrimer is $3 \times 2^4$ or 48 and the number for the fifth generation is $3 \times 2^5$ or 96. A fifth-generation dendrimer is shown in Figure 5.5.

There are two important aspects of dendrimers' structure from a sampling and collection perspective. The first is the voids or cavities in the interior of the structure and the second is the large number of functional groups on the surface of the structure. The voids can hold other molecules. The nature of the molecules will depend on the functional groups found in the interior of the dendrimer and the size of the voids. The surface groups can interact with other molecules based on covalent or hydrogen bonding type interactions. The surface groups can also be derivatized (functionalized). This results in the introduction of specificity.

### 5.5.5 Polymer Nanofibres

There are many examples of living organisms using fibre-based structures to remove solids from aqueous or airborne samples. Prawns, mentioned earlier in this chapter, and Baleen whales are examples. Polymer nanofibres are now commonly used in filter applications where there is a requirement to separate particulate matter from airborne or aqueous samples. Nanofibres have diameters less than 0.5 µm and are characterized by high surface-area-to-volume ratios.

Electrospinning is a common technique used to prepare nanofibres. The technique involves the application of a strong electric field to a polymer solution or a melted polymer. The solution or melt is fed through a small orifice at the tip of a die. The counter electrode is generally located several centimetres away and consists of a metal collection screen. The charging of the fluid or melt overcomes the surface tension of the droplet formed at the die tip and results in the solution or melt being ejected from the die tip and towards the counter electrode. The solution or melt thins rapidly due to elongation and evaporation of solvent and the resulting fibres are deposited on the counter electrode.[21]

Nanofibres are used in filtration units where they are very effective at removing particulate matter from airborne samples. They are also used to enhance the effectiveness of other filter media as they have little effect on the permeability of the filtration system. In addition, electrospun fibres increase the lifetime of these systems as they prevent or slow down the movement of particulate deeper into the underlying filter medium. They have also been incorporated into fabrics used in military uniforms to enhance resistance to the permeation of chemical or biological agents.

Fibres can also be co-electrospun to form core shell fibres. In this process two liquids of different chemical composition are electrospun from a nozzle (tip) with a core and a surrounding annular ring. One liquid flows through the core and the other through the surrounding annular ring and forms a compound droplet that is then electrospun. This process can be quick enough to prevent any mixing of the two liquids and can be used to form tubes if the core polymer can be removed using differences in chemical solubility or thermal stability. Electrospun fibres are also used as templates for the formation of nanofibre tubes (tubes by fibre templates (TUFT) process and as templates for wetting assisted templating (WASTE)).[22]

Derivatization of the surface of nanofibres or the interior and exterior surfaces of nanotubes allows the introduction of selectivity to these materials and enhances the ability of nanotubes to trap analytes of interest.

## 5.6 Bio-inspired/Biomimetic Collection and Sampling Systems

### 5.6.1 Biomimetic Air Sampling

The silkworm moth is capable of detecting very low concentrations of molecules without the use of any mechanical moving parts. The moth senses molecules *via* its antennae. An antenna has approximately 100 branches and each of the branches about 200 sensory hairs. The hairs, 2 μm wide and 100 μm long, contain about 3000 pores.[23] It is through these pores that molecules enter an aqueous fluid and are subsequently detected. This natural system can detect molecules at concentrations as low as 1000 molecules per cubic centimetre ($cm^3$).

Using a Monte Carlo technique, Futrelle[24] studied the motion of molecules when they enter an area covered with many small, post-like structures to understand the efficiency of the collection and detection process exhibited by the silkworm moth. The results were not consistent with the molecules sticking to the hairs of the antennae and then diffusing to the pores where they are detected. The rate of diffusion of the molecules through a liquid layer to the pores was orders of magnitude too slow to explain the observed response. Instead, it was concluded that the molecules were reflected off hairs (up to 250 times in a very short time) until they hit one of the many pores on the hairs. The molecule then passed through the pore and was detected.

Hughes[25] has investigated the use of preconcentration of analyte molecules to increase sensitivity. Polymeric materials were used to absorb analyte molecules. The absorbed molecules were rapidly released by heating and detected using a chemiresistor. Although this decreased detection limits substantially, the author suggests that coating posts that mimic the hairs of the silk moth antenna will further increase the ability of the coating to extract volatile organic compounds from air samples.

### 5.6.2 Water Collection and Transport (Thorny Devil)

The ability to collect and sample target material is not only crucial for detection and identification systems but can be critical for animals in certain harsh environments. All animals need water to survive. In dry environments water is scarce and the ability to find and use available water is critical for the survival of a species. The Australian Thorny Devil (*Moloch horridus*), shown in Figure 5.6(a), can take water from damp sand that it is standing on and move it rapidly through interscalar spaces on its skin's surface to its mouth for drinking.[26] The interscalar spaces are located in scale hinges. The hinged joint walls are covered by a complex topography of fractured surfaces that greatly expand the channels' surface area and probably enhance capillary transport of water. This system is so effective that *M. horridus* is capable of removing water from damp sand (3% water) by rubbing its ventral scales into it. This remarkable capability is illustrated in Figure 5.6(b) where blue dye has been used to show the movement of water from the foot pad to the head region of the Thorny Devil.

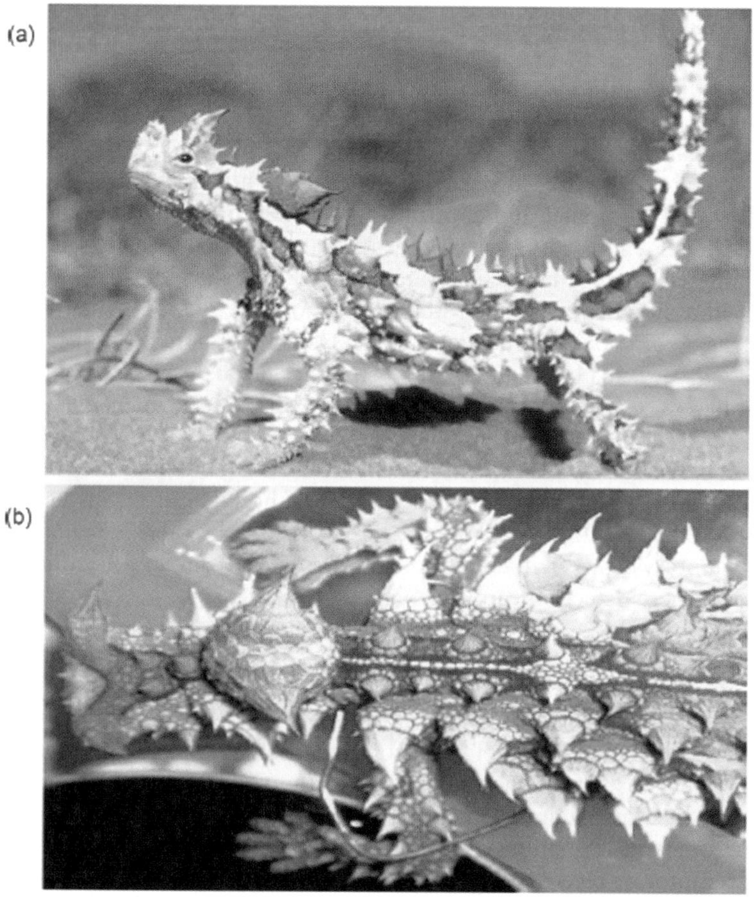
(a)
(b)

**Figure 5.6** (a) The Australian Thorny Devil (*Moloch horridus*). The reptile has evolved the ability to collect water from the environment through capillary action using its scales. (Photo courtesy of Paddy Ryan.) (b) Photograph showing the uptake of blue dye by *M. horridus*. This illustrates the mechanism used to collect water. (Photo courtesy of Lin Schwarzkopf, James Cook University, Australia.)

It is possible that studying these skin structures that have evolved to optimize water collection and transport may inspire new approaches to collection and sampling of chemical and biological agents.

### 5.6.3 Optimized/Controlled Fluid Flow

The vascular system in humans is a hierarchical structure where the diameter and length of blood vessels branching from another have evolved to minimize the work required to enable blood flow. In 1926, C. D. Murray[27] derived a relationship between the diameter of a blood vessel and the optimum diameters of blood vessels branching from it. This relationship, known as Murray's law, states that the sum of the cubes of the diameters of the smaller blood vessels should equal the cube of the diameter of the larger (parent) blood vessel.

This relationship is now being exploited to design microfluidic distribution systems.[28] The authors studied microfluidic manifolds with constant-depth rectangular or trapezoidal-sectioned channels that can be fabricated using microfabrication techniques such as photolithography and wet or dry etching. Numerical modelling was also carried out to show how the biomimetics theory could be used to predict and control flow through microfluidic channels. The numerical modelling was based on a generalization of Murray's law.[29,30] A branching parameter that governs the change in the diameter of channels at a branch in a system was defined in Murray's original paper. When this parameter has a value of one, the work in the system is minimized. However, it is possible to vary this branching parameter. This can be used to design systems with different hydrodynamic properties and introduce an element of flow control into the microfluidic system.

Water collection by the Thorny Devil and the design of fluid networks based on the human vascular system are examples of how Nature transports fluids. Knowledge of how these natural systems work can be (and is being) used to design fluid systems that can transport analytes from a collection site to receptors where detection can take place.

## 5.7 Conclusion

A major challenge in developing sensors incorporating bio-inspired collection and sampling systems is the integration of the collection and sampling capabilities with transduction elements. Even if a chemical can be sampled from the environment, it is still necessary that the chemical be brought into contact with a receptor or receptors so that detection can be achieved. The response to the collected analyte must also be processed. The integration of these capabilities in an olfactory system is fundamental to the ability of animals and insects to detect and differentiate between scent molecules.

The requirement for sampling and collection can also be affected by the specificity of the interaction between the receptor and the compound being sampled. When the interaction between a compound and a receptor is specific, this can reduce or eliminate the requirement for collection and preconcentration prior to detection. It is also possible to have the sampling/receptor portion of a sensing system in direct contact with the transducer, which would eliminate the need to transfer the compound from the area where it was sampled to the area where it is detected.

A major challenge to the use of synthetic receptors, such as MIPs, for sensing applications is to prepare them in forms other than powders or thin films. The ability to prepare two-dimensional (2-D) and three-dimensional (3-D) structures is still in its developmental stages. A number of techniques including lithography, microcontact printing, microstereolithography and 2-photon 3-D lithography have been used to produce the type of 2-D and 3-D structures required for sensor applications. A microstereolithography technique has been described in the literature[31] for the preparation of 2-D and 3-D structures. Localized photopolymerization with a focused laser beam was used to build the MIP structure.

Collection and sampling by natural (biological) systems or materials can also have limitations. These systems may be susceptible to extreme temperatures, variations in pH and relative humidity, and have incompatibilities with certain chemicals. The materials used for sampling and collection may also lack the mechanical and chemical properties that are required for certain in-service environments. There are also instances where no natural material exists with the properties required to sample and collect a particular compound. Development of materials that address these limitations while keeping the desirable bio-inspired properties, such as specificity, will address some of these limitations.

## References


1. L. Turin, *Chem. Senses*, 1996, **21**, 773.
2. L. Turin, *J. Theor. Biol.*, 2002, **216**, 367.
3. Science Daily, 15th September, 2009, http://www.sciencedaily.com/releases/2009/02/090213172533.htm, accessed 6th October, 2009.
4. F. Rock, N. Barsan and U. Weimar, *Chem. Rev.*, 2008, **108**, 705.
5. M. A. Sánchez-Montañés, J. W. Gardner and T. C. Pearce, *Proc. R. Soc. A*, 2008, **464**, 1057.
6. J. W. Gardner, J. A. Covington, S.-L. Tan and T. C. Pearce, *Proc. R. Soc. A*, 2007, **463**, 1713.
7. D. Kris, O. Ramstrom and K. Mosbach, *Anal. Chem.*, 1997, **345A**.
8. R. A. Potyrailo, *Angew. Chem., Int. Ed.*, 2006, **45**, 702.
9. K. Mosbach and O. Ranstrom, *Bio/Technology*, 1996, **14**, 163.
10. G. Wulff, *Angew. Chem. Int. Ed. Engl.*, 1995, **34**, 1812.
11. S. A. Piletsky and A. P. F. Turner, *Electroanalysis*, 2003, **14**, 317.
12. M. E. Davis, *Nature*, 2002, **417**, 813.
13. C. Baerlocher and L. B. McCusker, International Zeolite Association, 2009, vol. 2009.

14. M. M. J. Treacy and J. B. Higgins, *Collection of Simulated XRD Powder Patterns for Zeolites*, Elsevier Science and Technology Books, 5th revised edn, 2007.
15. R. A. Dunbar, J. D. Jordan and F. V. Bright, *Anal. Chem.*, 1996, **68**, 604.
16. T. A. Dickenson, J. White, J. S. Kauer and D. R. Walt, *Nature*, 1996, **382**, 697.
17. K. J. Shea and D. A. Loy, *MRS Bulletin*, 2001, May, 368.
18. C. J. Brinker, R. Seghal, S. L. Hietala, R. Deshpande, D. M. Smith and C. S. Ashley, *J. Membr. Sci.*, 1994, **85**, 94.
19. A. Katz and M. E. Lewis, *Nature*, 2000, **403**, 6767.
20. S. Svenson and D. A. Tomalia, *Adv. Drug Deliv. Rev.*, 2005, **57**, 2106.
21. Z. -M. Huang, Y. -Z. Zhang, M. Kotaki and S. Ramakrishna, *Compos. Sci. Tech.*, 2003, **63**, 2223.
22. R. Dersch, M. Steinhart, U. Boudriot, A. Greiner and J. H. Wendorff, *Polym. Adv. Technol.*, 2005, **16**, 276.
23. R. C. Hughes, December 2003, SAND Report, SAND 2003, 4707.
24. R. P. Futrelle, *Trends Neurosci.*, 1984, **7**, 116.
25. R. C. Hughes, Biomimetic Air Sampling for Detection of Low Concentrations of Molecules and Bioagents, LDRD 52744 Final Report, Sandia National Laboratories, 2003.
26. W. C. Sherbrooke, A. J. Scardino, R. D. Nys and L. Schwarzkopf, *Zoomorphology*, 2007, **126**, 89.
27. C. D. Murray, *Proc. Natl. Acad. Sci. USA*, 1926, **12**, 207.
28. R. W. Barber, K. Cieslicki and D. R. Emerson, in *Design and Nature III: Comparing Design in Nature with Science and Engineering*, WIT Press, Southampton, UK, 2006, p. 245.
29. K. Cieslicki, *Polish J. Med. Phys. Eng.*, 1999, **5**, 161.
30. R. W. Barber and D. R. Emerson, *Microfluid Nanofluid*, 2008, **4**, 179.
31. P. G. Conrad, P. T. Nishimura, D. Aherne, B. J. Schwartz, D. M. Wu, N. Fang, X. Zhang, J. Roberts and K. J. Shea, *Adv. Mater.*, 2003, **15**, 1541.

CHAPTER 6
# Receptors and Surfaces

## 6.1 Introduction

Receptors and surfaces represent the starting point for every sensor or sensing system. This recognition or binding element may take many forms, but in the case of bio-inspired technologies it may encompass natural receptors, *i.e.* antibodies, whole cells and isolated DNA, or utilize artificial receptors synthesized using biological approaches, for example molecularly imprinted polymers (MIPs) and functionalized surfaces on virus nanoparticles. Though there are numerous comprehensive reviews and books describing receptors and biosensors in great detail,[1–6] the technologies described in this chapter are intended to highlight bio-derived or biomimetic receptors and surfaces that show particular promise for defence and security. As suggested by Iqbal *et al.*,[7] improvements in the affinity, specificity and mass production of the molecular recognition components may ultimately dictate the success or failure of detection technologies in both a technical and commercial sense. These authors also predicted that the vision of giving the individual soldier on the battlefield or civilian responders to a biological attack or epidemic, a miniature, sensitive and accurate biosensor may depend as much on molecular biology and molecular engineering as on hardware engineering.[7]

Two general approaches have emerged in the development of receptors and surfaces for detection. Recognition elements already found in Nature can be isolated for use in sensors or natural receptors can be duplicated using surfaces or synthetic materials based on the molecular structures in a biomimetic fashion. As illustrated in Figure 6.1, receptors most commonly used in sensor applications can be further divided into four general areas: antibodies, molecular bioprobes, synthetic ligands and functional surfaces.

---

Bio-inspired Materials and Sensing Systems
By Peter Biggins, John Hiltz and Anne Kusterbeck

Published by the Royal Society of Chemistry, www.rsc.org

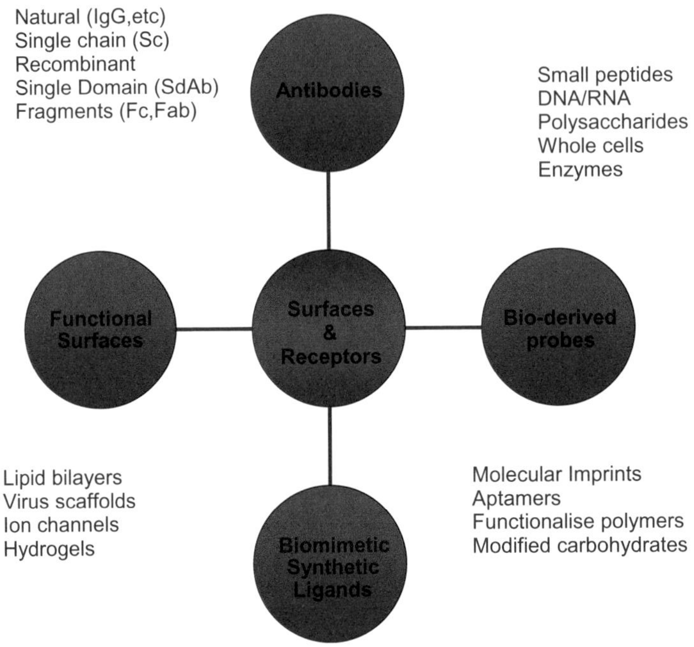
Natural (IgG,etc)
Single chain (Sc)
Recombinant
Single Domain (SdAb)
Fragments (Fc,Fab)
Antibodies
Small peptides
DNA/RNA
Polysaccharides
Whole cells
Enzymes
Functional Surfaces
Surfaces & Receptors
Bio-derived probes
Lipid bilayers
Virus scaffolds
Ion channels
Hydrogels
Biomimetic Synthetic Ligands
Molecular Imprints
Aptamers
Functionalise polymers
Modified carbohydrates

**Figure 6.1** Detailed analysis of receptors and surfaces for sensor development.

## 6.2 Natural Receptors

In looking at the natural receptors, the classic model and diagnostic 'work horse' for detection are antibodies. Other bio-derived binders including DNA and protein probes are also in widespread use.

### 6.2.1 Antibodies

Antibodies (Abs) stand in a class by themselves with respect to bio-derived recognition elements. The acceptance and broad popularity of antibodies is due primarily to the well-known sensitivity and specificity of these proteins as well as their long-standing use in the medical community as key components in detecting pathogens and disease and the defence-related use of antibody-based assays for biothreat detection, both in the laboratory and in the field (reviewed in Dover *et al.*,[3] Byrne *et al.*,[8] and Conroy *et al.*[9]). The most common format for detection is the immunoassay, in which Abs produced by an organism as an immune response are isolated and characterized for specificity, binding affinity and selectivity for a given target, such as a biological or chemical agent. From a defence perspective, antibodies in their various forms are still the basis for most conventional biological threat detection scenarios. Figure 6.2(a) illustrates the many constructs these receptors may take when used as part of a sensor and Figure 6.2(b) provides a schematic of the various formats commonly employed when developing an antibody-based detection system.

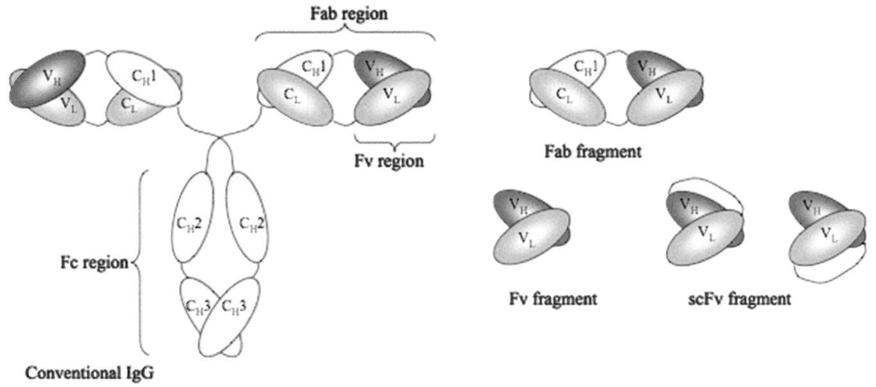
Fab region
Fv region
Fab fragment
Fc region
Fv fragment
scFv fragment
Conventional IgG

(a)

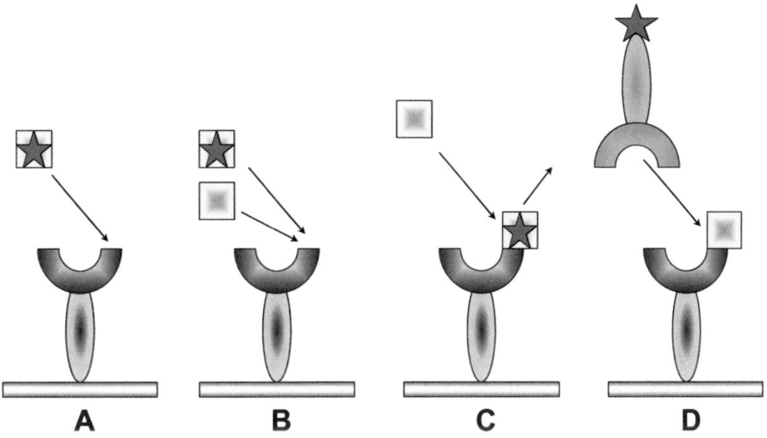
A
B
C
D

(b)

**Figure 6.2** a) Structure of native IgG antibody and antibody fragments with isolated binding regions typically used in immunoassays. Accessed from http://www.microbialcellfactories.com/content/2/1/1 © 2003 Joosten *et al.*; licensee BioMed Central Ltd. This is an Open Access article: verbatim copying and redistribution of this article are permitted in all media for any purpose, provided this notice is preserved along with the article's original URL. b) Illustration of typical formats used for antibody-based assays with immobilized antibodies (blue molecule), antigens (yellow square) and labels (red star). A Direct binding immunoassay, B Competitive immunoassay, C Displacement immunoassay, D Sandwich immunoassay with second antibody.

### *6.2.1.1 Immunoglobulin G (IgGs)*

These large proteins are used in the majority of assays and initial sensor development due to the many well-tested protocols and platforms that have been developed using IgGs. As seen in Figure 6.2, the intact IgG molecule has two binding regions (epitopes) that can act as receptors for the targets of choice. While IgGs isolated from immunized animals (polyclonal Abs) have a wide range of binding affinities, the later development of monoclonal antibodies from mouse hydridoma cells enabled the production of high quantities of antibodies with well-characterized specificity and binding at a single epitope. As demonstrated by multiple investigators, these basic forms of the antibody are extremely effective in detecting biothreat agents, bacteria, toxins and small molecular weight chemicals.[10–12]

#### *6.2.1.2 Antibody Fragments*

While whole antibodies remain the receptor of choice in assays and sensor development, a subcategory of antibodies known as antibody fragments and peptide probes is rapidly gaining ground due to the ease with which they can be produced (reviewed in Weisser and Hall[13]). Generated using recombinant antibody technologies, these partial segments of antibodies consist primarily of the isolated variable binding region separated from an intact antibody by various proteolytic cleavage protocols (see Figure 6.2(a)) for detailed structures). The primary fragments used are:

1) Fab and $F(ab')_2$ – For these constructs, the two variable binding regions of the antibody are isolated along with the constant regions to maintain stability.
2) scFv – In these receptors, only the single chain variable region fragment (scFv) portion of the Ab is cleaved from the IgG molecule. In general, the most effective way of using these fragments is by incorporating flexible peptide linkers to serve as tethers to a surface and assist in retaining binding functionality.

#### *6.2.1.3 Single Domain Antibodies (sdAbs)*

Camelids and sharks have unique immune systems that produce antibodies with binding sites formed from a single peptide chain, as opposed to conventional antibodies (*i.e.* those found in humans, mice and rabbits) that require two separate chains (Figure 6.3). An additional benefit to these receptors is their ability to retain binding activity even after prolonged exposure to high temperatures, demonstrated in Figure 6.3 with the assay for ricin in which the sdAb remained active despite being held at 95 °C for 30 minutes. Finally, routine techniques in molecular biology can be readily applied to isolate, screen and engineer the binding properties of these small molecules using affinity maturation. Researchers, including Goldman *et al.*,[14,16] and Hayhurst *et al.*,[15] are now actively working to isolate the active binding regions of these sdAbs to produce large quantities of receptors specific for biothreat agents. They have demonstrated assays for ricin and botulinum toxin as well as Brucellis.[14–16]

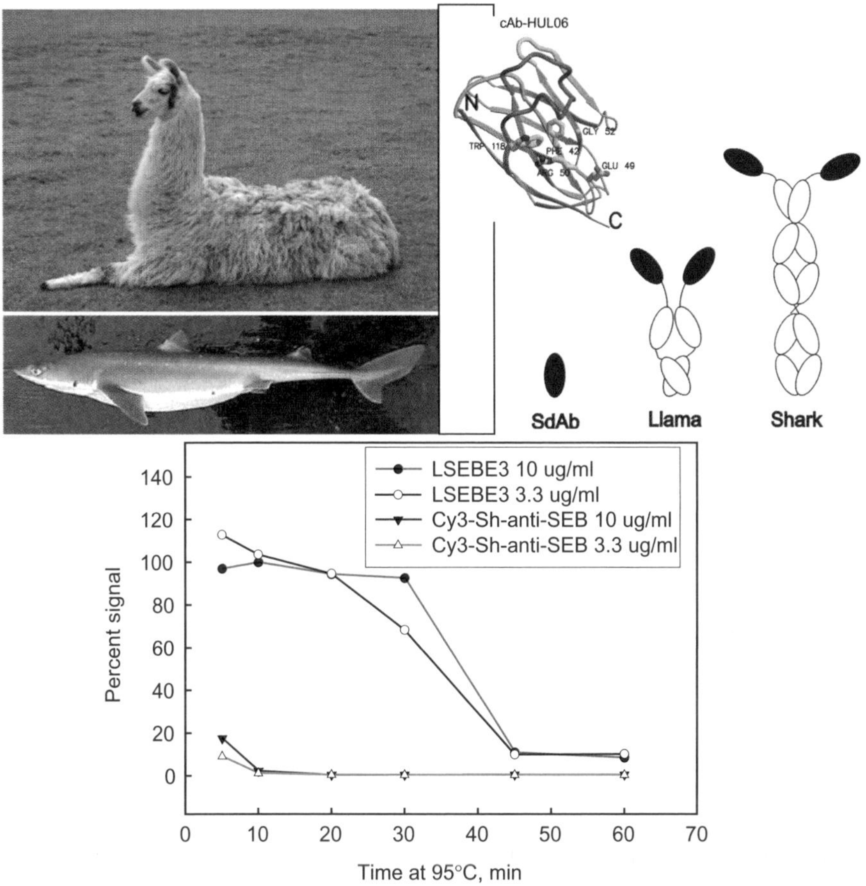
cAb-HUL06
N
C
SdAb
Llama
Shark
LSEBE3 10 ug/ml
LSEBE3 3.3 ug/ml
Cy3-Sh-anti-SEB 10 ug/ml
Cy3-Sh-anti-SEB 3.3 ug/ml
Percent signal
0
20
40
60
80
100
120
140
0
10
20
30
40
50
60
70
Time at 95°C, min

**Figure 6.3** Llama and shark single domain antibodies (sdAb) as receptors in immunoassays with improved thermal stability. Shown here are the structures of the intact antibodies and the single domain fragment (top) and an assay for staphylococcal enterotoxin B (SEB) (bottom) in which activity of the llama sdAb, LSEBE3, is maintained for over 30 min at high heat as compared to traditional IgGs (anti-SEB).

## 6.2.2 Other Bio-derived Molecular Bioprobes

While antibodies remain the predominant receptor, molecular recognition using DNA-/RNA-based methods is becoming more widespread due to the advances in genomics and sequencing technologies.

### *6.2.2.1 Nucleic Acids*

A complex system of recognition, binding and response already exists in the nucleic acid genetic material, DNA and RNA. This type of recognition is due to the base pair matching of one strand of a nucleic acid to its opposite, complementary strand. Any pathogen with DNA or RNA can be detected using this method but it cannot normally be used with toxins or other chemicals. Nucleic acid probes have advanced rapidly since the early development of the polymerase chain reaction (PCR) methods in which multiple copies of the genetic material are produced through rapid cycles of melting and rebinding. Lin and coworkers[17] have recently described a resequencing DNA microarray for broad spectrum pathogen identification that can be used as a diagnostic and surveillance tool to test for thousands of biological organisms in parallel. This new platform for biosurveillance uses human gene subtraction, random amplification of the pathogen DNA and high-density gene arrays combined with bioinformatics to identify pathogens at the strain level. Initial clinical studies on patient samples demonstrated 99% sensitivity with less than 1% false positive.[17,18]

#### 6.2.2.2 *Antimicrobial Peptides*

While antibodies provide specificity in assays, a more generic, broad-based detection of biothreat agents may be accomplished using the natural receptors provided by antimicrobial peptides (AMPs). This promising class of generic receptors has long been known to be part of the innate immunity possessed by many organisms to defend against bacteria. Recently these peptides have been isolated and characterized for their response to biothreat agents. AMPs are alternative recognition elements that are thought to insert into the membrane of bacterial cell walls and disrupt them, causing cell death. Different forms of these AMPs have been isolated from the skin of frogs and used as antibiotics. When used as a receptor, they bind multiple targets with overlapping specificities and differing affinities as demonstrated by Kulagina *et al.*,[19] and Arcidiacono *et al.*[20] Binding is semi-selective and a target-specific reagent is not required, allowing AMPs to be used in single, all-encompassing tests for bacteria, toxins, viruses and fungi, including biothreat agents like rickettsia.[19] In addition, these peptides are heat- and salt-stable, making them attractive for field use. Shown in Figure 6.4 is the structure of one of these AMPs isolated from frogs.

#### 6.2.2.3 *Cell-based Sensing*

In the case of cell-based sensing, the entire cell itself with multiple receptors expressed on the surface is used to bind the analytes of interest, thus eliminating the need for isolating and immobilizing functional receptors. With all its inherent complex responses and feedback loops intact, this type of receptor also offers an ideal generic, broad-based detection mechanism for a variety of targets. These may include toxins, bacteria and viruses. An additional benefit to cell-based systems is that the physiological effect of the targets can be measured and evaluated.[21,22] Modern methods of cell culture, high-throughput screening,

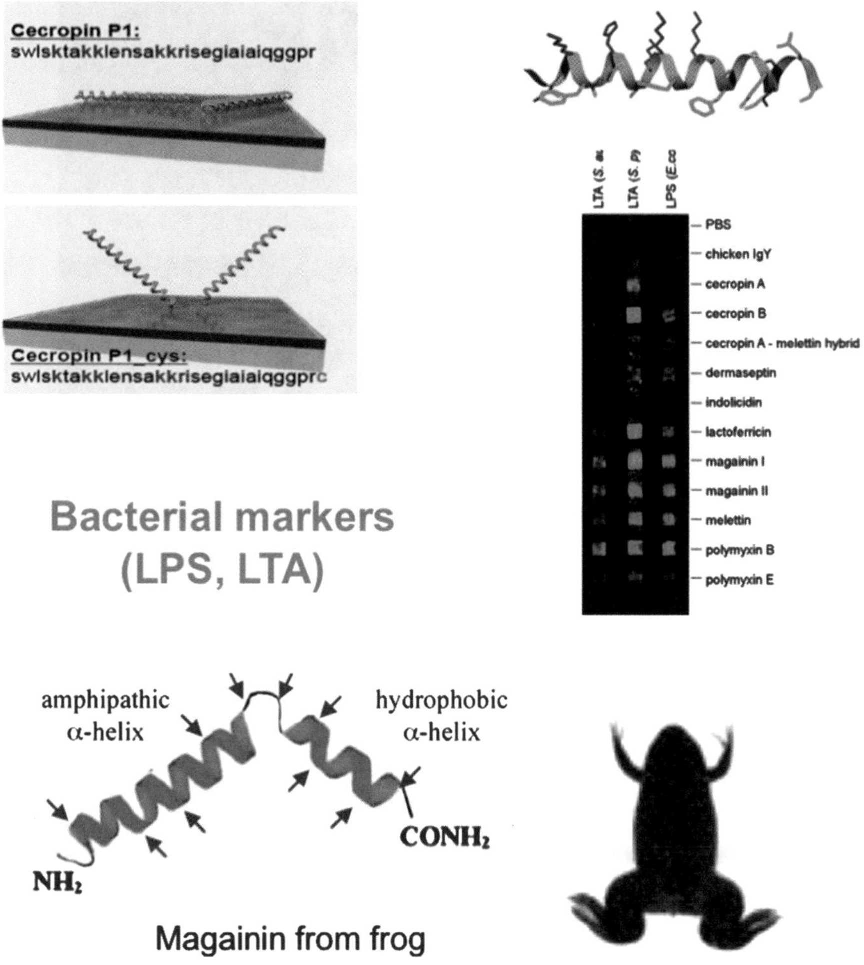
Cecropin P1:
swlsktakklensakkrisegiaiaiqggpr
Cecropin P1_cys:
swlsktakklensakkrisegiaiaiqggprc
Bacterial markers
(LPS, LTA)
LTA (S. au
LTA (S. P)
LPS (E.co
PBS
chicken IgY
cecropin A
cecropin B
cecropin A - melettin hybrid
dermaseptin
indolicidin
lactoferricin
magainin I
magainin II
melettin
polymyxin B
polymyxin E
amphipathic
α-helix
hydrophobic
α-helix
NH2
CONH2
Magainin from frog

**Figure 6.4** Schematic of antimicrobial peptides isolated from frogs used as receptors for generic detection of biothreat agents.

micro fluidics and the ability to analyse signalling events with electronics have increased the utility of cellular methods (for a review, see Banerjee and Bhunia[2]). Figure 6.5 illustrates how one such system has been developed by patterning mammalian neuronal cells on microelectrode arrays (MEAs) so that exposure to a toxin can be detected by the change in electrical activity of the cell itself.[23,24] O'Shaughnessy *et al.*, have further expanded this capability by imparting functionality to neuronal stem cells, thus avoiding previous problems with finite cell life spans.[25]

Taking this concept further, molecular biology and genetic engineering techniques have provided the ability to make cells express receptors for specific analytes.[26,27] This approach has also led to commercial instruments discussed

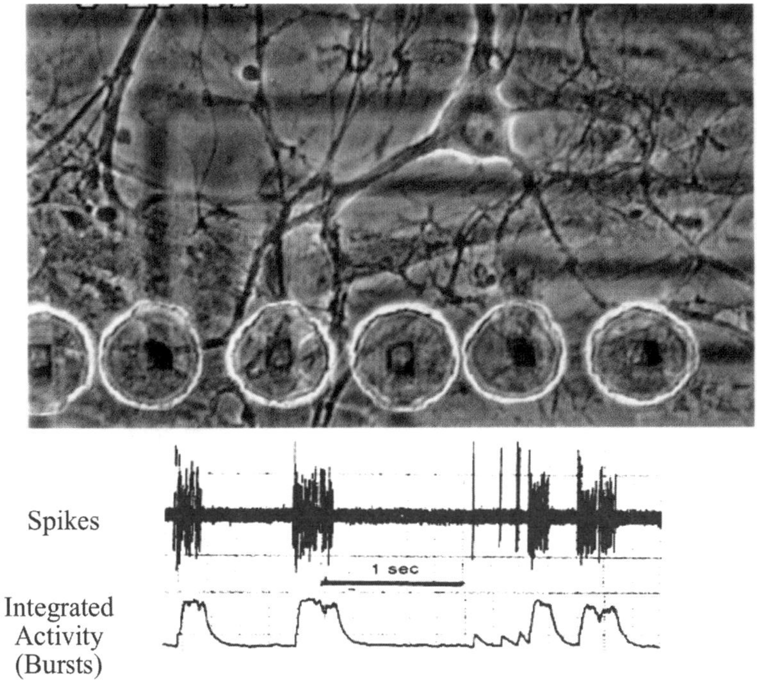
Spikes
1 sec
Integrated
Activity
(Bursts)

**Figure 6.5** Neuronal cell-based sensor detection of toxins. Shown are cells grown on MEA's and the response recorded upon exposure to toxins.

in Chapter 1 and the potential application of these 'canary' technologies in biodefence scenarios by first responders.[2]

### 6.2.3 Synthetic Ligands

The natural receptors and probes discussed above are isolated from organisms and have innate binding of various levels. For realistic use of the probes in a conceptual SASS, the area of receptor development that is of most interest to the defence community is that of synthetic ligands.[28] Because most operational environments are difficult and uncertain, the most desirable sensor components would be those that can withstand harsh conditions, including high temperatures, varying humidity and rapid recycling or regeneration. The recognition elements highlighted in this section meet many of those criteria.

#### *6.2.3.1 Aptamers*

While the Ab fragments discussed in a previous section are natural products based on immunizing animals or are developed using cell culture methods, aptamers are biomimetic artificial nucleic acid (DNA/RNA) ligands, also known as 'chemical antibodies', since they can be produced *in vitro* against targets that would otherwise be toxic to living organisms.[29] Aptamers are synthesized through multiple cycles of selection and amplification, known as Systematic Evolution of Ligands by Exponential Enrichment (SELEX). This chemical technique for generating combinatorial libraries of the aptamers produces receptors that bind non-nucleic acid targets with both high affinity and high specificity. Unlike the complementary base pairing of DNA probes, their method of binding is based on the three-dimensional interaction of the synthetic nucleic acids with proteins (see Tombelli *et al.* for a review[30,31]). A key advantage of aptamers is that the single strands can be chosen for a given target, beginning with screening against a large library ($10^{15}$ or more) of possible sequences. In addition, since the ligands are synthetic, they can be developed against material that would normally be toxic to cells or animals if used in antibody production.

#### *6.2.3.2 Molecularly Imprinted Polymers*

These supramolecular, biomimetic materials were discussed extensively in Chapter 5 with a focus on their use as a material for sample collection and preconcentration. Perhaps of more importance, however, is the role MIPs may play as rationally designed artificial receptors with hierarchical structures that can be engineered to bind small molecules while withstanding extreme conditions, the so-called synthetic antibodies (see Piletsky and Turner[32] for a review). Similar to the benefits of aptamers, the MIPs can be produced in large quantities, are resistant to extreme temperatures and can be tuned with regard to porosity, surface area and other material properties. They are also compatible with microfabrication methods and can be made against a broad range of analytes. As receptors in sensor development, the major limitations include poor batch-to-batch reproducibility, low binding affinity and specificity and poor performance in aqueous environments. Their primary strength is in the detection of low molecular weight chemicals such as the explosive, trinitrotoluene (TNT),[33] pyrene and volatile organic compounds.[34] Efforts are continuing to identify and improve MIPs or other synthetic binders to overcome current problems with weak binding affinity and specificity.

## 6.3 Functionalized Surfaces

Though receptors have received the majority of the focus when developing sensors, surfaces also play an important role in providing functionality. The key features that should be considered include elasticity, specificity, structure and compatibility with the surrounding environment. The ability to mimic biological surfaces has also been studied extensively since these structures allow a more realistic interface with biological target analytes such as biothreat agents. Bio-inspired or biomimetic surfaces are being developed that have a broad range of functions and properties that can provide a simple substrate for receptors as well as other desirable characteristics needed at a systems level,

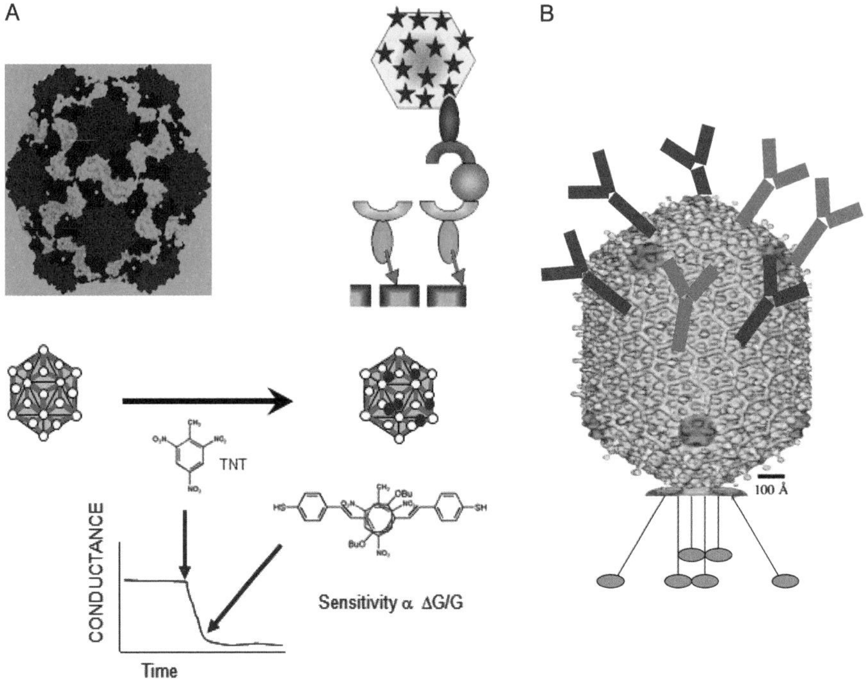
A
B
TNT
CONDUCTANCE
Time
Sensitivity α ΔG/G
100 Å

**Figure 6.6** Viral particles as A) scaffolds to immobilize antibodies for an immunoassay and B) mechanism for phage expression of antibodies on the T4 Tobacco Mosaic Virus (TMV) head.

including chemical resistance or compatibility (hydrophobicity and hydrophilicity), camouflage, protection of substrates and selective permeability. This section will present an overview of current surfaces being utilized in sensor development.

### 6.3.1 Virus Particles as Scaffolds

A unique example of bio-inspired technology is the use of the surface and structure of virus particles as a scaffold. The virus particles offer the advantage of high surface areas, well-defined spatial control of the surface features and morphology and multiple sites for attachment of labels. In addition, large-scale production and genetic manipulation of the particles is possible and they are stable in harsh environments. Cowpea mosaic virus (CMV), a natural plant virus, has multiple, well-defined functional groups that can be used to attach ligands, such as antibodies, to their surface for assembly into sensing systems.[35] Shown in Figure 6.6(a) is an example of how the CMV particle can be modified chemically through site-directed mutagenesis to produce functionalized surfaces for attachment of proteins. Figure 6.6(b) shows a schematic of a similar approach using the T4 bacteriophage in which the phage itself is genetically induced to express multiple antibodies, allowing the expression of multiple receptors.

### 6.3.2 Lipid Bilayers

These surfaces represent the classic model of biomimetic systems in that they seek to impart improved functionality to sensors by setting up an environment that is more like a cell wall or bilayer membrane. They also could be considered to be the basic structure for sensing in organisms since receptors in Nature do not exist in isolation but are anchored in the lipid membrane of the cell. This relationship provides stability to the binding element as well as contributing to the transduction event by allowing transport of ions across the membrane. As envisioned by Liebeau *et al.*,[36] the self-assembled bilayer containing glycolipids is an ideal surface for the interaction of carbohydrates, cells and toxins. To mimic more closely this natural process, various groups have successfully demonstrated the deposition of lipid bilayers on polymers,[37] hydrogels[38] and porous materials.[39] Comprehensive reviews have been done on these artificial membranes highlighting preparation methods, lipid composition and attachment mechanisms to immobilize the bilayer (see Janshoff and Steinem,[40] Tien *et al.*, 2001[41] and Hirano-Iwata *et al.*[39]). When combined with new methods to deposit the films on a support or micro-machined substrates, detection can be accomplished very rapidly as compared to traditional methods, especially when electrochemical transduction is used. The current substrates of choice are gold and silica, based largely on their ease of use and compatibility with transducers.

Another promising concept being explored is that of tethered lipid bilayers. This technology combines the functional cell membrane platform with nanopatterned hydrogels or polymers.[37,38,42] Self Assembled Monolayer (SAM) surfaces patterned with nanostructures that recognize phenyl groups have also been developed to mimic olfaction.[43] Rather than bind a specific compound, these surfaces bind molecular substructures of a benzene ring, recognizing the hydrophobicity and the molecular size.

### 6.3.3 Hydrogels

A final functionalized surface that is being used on multiple sensing platforms is hydrogels. Several groups have reported 3-D protein hydrogels with cross-linked polymers[44] that can be used to immobilize other functional groups or cells. Such functionalized systems have much higher surface area, thereby increasing sensitivity and response.[45] Others are developing 'bio-smart' hydrogels that incorporate both the receptor and transducer[46] or using the surface to provide structure for protein microarrays.[47] An example of hydrogel structures used to immobilize cells and proteins is shown in Figure 6.7.

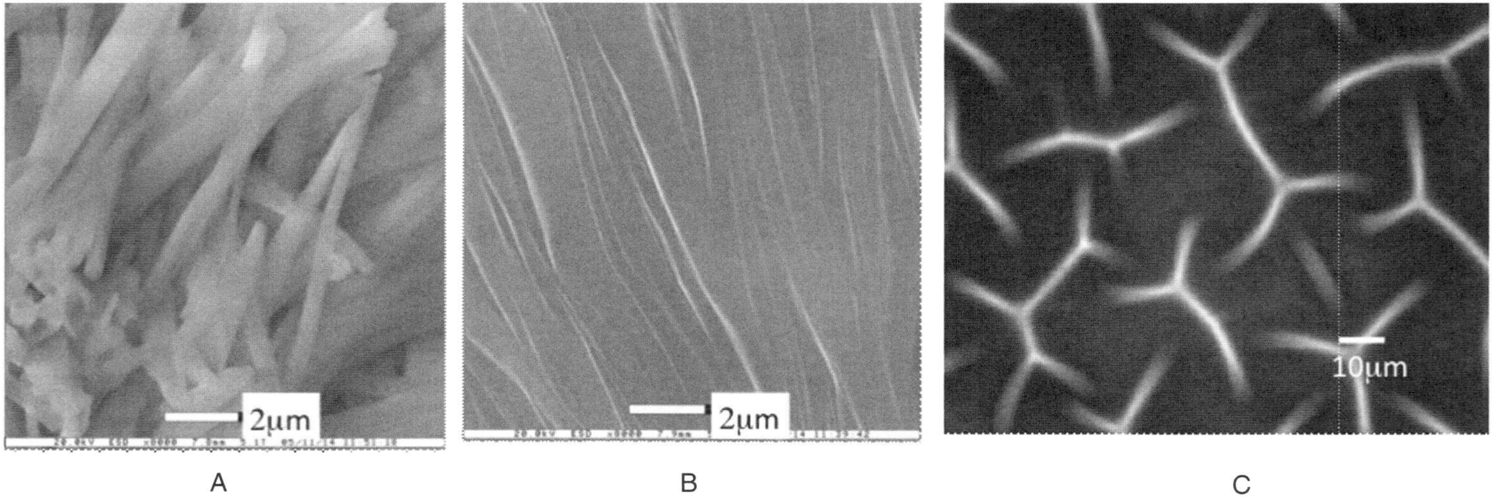
2µm
2µm
10µm
A
B
C

**Figure 6.7** Fine structure of hydrogels, an example of functionalized surfaces used in sensor development. Shown are (A) SEM images at low magnification, (B) higher magnification and (C) a light microscopy image.

### 6.3.4 Nanoarrays with Bio-inspired Nanocorals

A unique bio-inspired surface recently introduced is a type of array analogous to sea coral, which have an extremely rough external surface to increase surface area and improve the capture of food. These nanocoral sensing arrays are fabricated using polystyrene nanospheres on glass, with subsequent plasma etching that produces trenches and a coating with gold to allow surface-enhanced Raman spectroscopy (SERS).[48]

## 6.4 On the Horizon: Molecular Biomimetics

An exciting new area of receptor research that is a true multidisciplinary approach between physics, engineering, biology and chemistry is that of molecular biomimetics. As highlighted throughout this chapter, the recognition elements used in current technologies are based on molecular building blocks that span the range from nano- to microscale. If traditional physics and biological components could be used to form hybrid materials that bind like proteins, a true engineering approach to receptor development could be realized. In a similar fashion, directed molecular evolution, much like natural evolution of

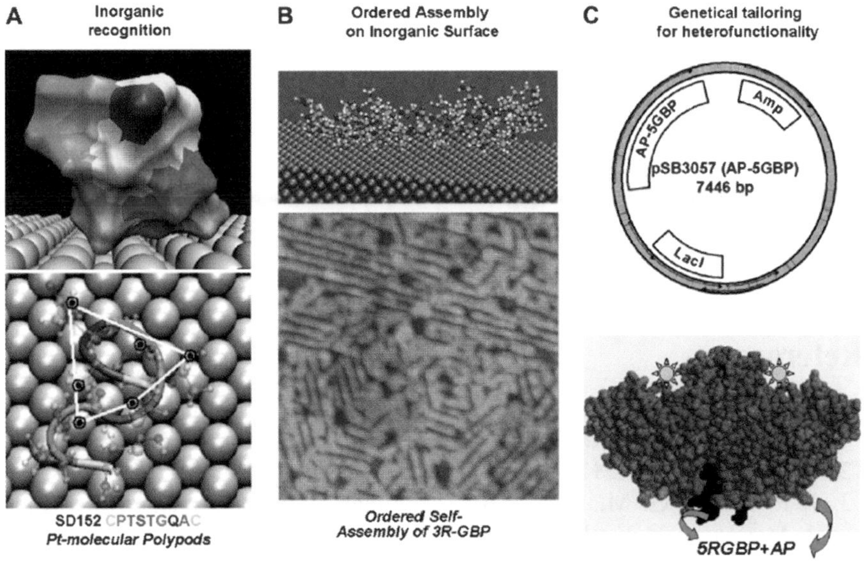
A
Inorganic recognition
B
Ordered Assembly on Inorganic Surface
C
Genetical tailoring for heterofunctionality
pSB3057 (AP-5GBP) 7446 bp
AP-5GBP
Amp
LacI
SD152 CPTSTGQAC
Pt-molecular Polypods
Ordered Self-Assembly of 3R-GBP
5RGBP+AP

**Figure 6.8** Illustration of how molecular biomimetics might be used to engineer specific receptors molecular recognition, self-assembly and genetic manipulation: (A) Recognition of platinum, a noble metal, by a phage-display selected Pt-binding. (B) Genetically designed three-repeat gold-binding peptide that recognizes the Au(111) surface and assembles into a two-dimensional ordered supramolecular structure. (C) Plasmid construct used to construct a bi-functional molecule (fom C. Tamerler and M. Sarikaya, *Acta Biomaterialia*, 2007, **3**, 289–299, with permission of Elsevier).

desirable characteristics, is emerging in which receptor structure and function are altered using combinatorial chemistry and molecular biology as a form of directed molecular evolution.[49] In all these cases, the vision is to develop components that can be interchangeably used as building blocks for the ideal molecular recognition element. As one example, a random library of peptides with the same number of amino acids, but varying compositions, is screened to identify specific sequences that strongly bind to an inorganic material of practical interest. In molecular biomimetics, this set of building blocks would be part of a molecular 'erector set' in which different proteins, each engineered to bind to a specific surface, size or morphology of an inorganic compound, promote the assembly of intricate, hybrid structures composed of inorganics, proteins and even functional polymers. As suggested by these researchers and illustrated in Figure 6.8, achieving this would be a giant leap toward realizing nanoscale building blocks in which the protein and its binding characteristics are tailored using DNA technologies, thereby creating the perfect receptor.[50]

## 6.5 Conclusion

Receptor technology has advanced rapidly in parallel with new discoveries in molecular biology and an improved basic understanding of how biological recognition elements can be modified to impart specific functionalities. Though the promise of the perfect synthetic antibody has yet to reach the exquisite sensitivity and selectivity of natural antibodies, there are a number of molecular receptors that are excellent binders for biothreat agents and chemicals, with the added advantage that they can be produced in large quantities at lower costs. The emerging receptors and surfaces can also be expected to withstand rugged field conditions that are required in fieldable systems, including an autonomous sensing system like the SASS.

## References

1. G. D. Griffin, D. N. Stratis-Cullum and S. Moselio, in *Encyclopedia of Microbiology, (Third Edition)*, Academic Press, Oxford, 2009, pp. 88–103.
2. P. Banerjee and A. K. Bhunia, *Trends Biotechnol.*, 2009, **27**, 179–188.
3. J. E. Dover, G. M. Hwang, E. H. Mullen, B. C. Prorok and S.-J. Suh, *J. Microbiol. Methods*, 2009, **78**, 10–19.
4. M. E. McConney, K. D. Anderson, L. L. Brott, R. R. Naik and V. V. Tsukruk, *Adv. Funct. Mater.*, 2009, **19**, 2527–2544.
5. D. V. Lim, J. M. Simpson, E. A. Kearns and M. F. Kramer, *Clin. Microbiol. Rev.*, 2005, **18**, 583.
6. F. S. Ligler, K. E. Sapsford, J. P. Golden, L. C. Shriver-Lake, C. R. Taitt, M. A. Dyer, S. Barone and C. J. Myatt, *Anal. Sci.*, 2007, **23**, 5–10.
7. S. S. Iqbal, M. W. Mayo, J. G. Bruno, B. V. Bronk, C. A. Batt and J. P. Chambers, *Biosens. Bioelectron.*, 2000, **15**, 549–578.

8. B. Byrne, E. Stack, N. Gilmartin and R. O'Kennedy, *Sensors*, 2009, **9**, 4407–4445.
9. P. J. Conroy, S. Hearty, P. Leonard and R. J. O'Kennedy, *Semin. Cell Dev. Biol.*, 2009, **20**, 10–26.
10. K. E. Sapsford, C. Bradburne, J. B. Delehanty and I. L. Medintz, *Mater. Today*, 2008, **11**, 38–49.
11. F. S. Ligler, in *Optical Chemical Sensors, Nato Science Series I: Mathematics Physics and Chemistry*, ed. F. Baldini, A. N. Chester, J. Homola and S. Martellucci, 2006, vol. 224, pp. 437–455.
12. C. R. Taitt, L. C. Shriver-Lake, M. M. Ngundi and F. S. Ligler, *Sensors*, 2008, **8**, 8361–8377.
13. N. E. Weisser and J. C. Hall, *Biotechnol. Adv.*, 2009, **27**, 502–520.
14. E. R. Goldman, J. L. Liu, R. D. Bernstein, M. D. Swain, S. Q. Mitchell and G. P. Anderson, *Sensors*, 2009, **9**, 542–555.
15. J. O. Conway, L. J. Sherwood, M. T. Collazo, J. A. Garza and A. Hayhurst, *Plos One*, 2010, **5**(1), E8818.
16. E. R. Goldman, G. P. Anderson, J. L. Liu, J. B. Delehanty, L. J. Sherwood, L. E. Osborn, L. B. Cummins and A. Hayhurst, *Anal. Chem.*, 2006, **78**, 8245–8255.
17. B. Lin, A. P. Malanoski, Z. Wang, K. M. Blaney, A. G. Ligler, R. K. Rowley, E. H. Hanson, E. von Rosenvinge, F. S. Ligler, A. W. Kusterbeck, D. Metzgar, C. P. Barrozo, K. L. Russell, C. Tibbetts, J. M. Schnur and D. A. Stenger, *Plos One*, 2007, **2**(5), e419.
18. C. R. Taitt, A. P. Malanoski, B. C. Lin, D. A. Stenger, F. S. Ligler, A. W. Kusterbeck, G. P. Anderson, S. E. Harmon, L. C. Shriver-Lake, S. K. Pollack, D. M. Lennon, F. Lobo-Menendez, Z. Wang and J. M. Schnur, *FEMS Immunol. Med. Microbiol.*, 2008, **54**, 356–364.
19. N. V. Kulagina, K. M. Shaffer, F. S. Ligler and C. R. Taitt, *Sens. Actuator, B Chem.*, 2007, **121**, 150–157.
20. S. Arcidiacono, P. Pivarnik, C. M. Mello and A. Senecal, *Biosens. Bioelectron.*, 2008, **23**, 1721–1727.
21. S. A. Gray, J. K. Kusel, K. M. Shaffer, Y. S. Shubin, D. A. Stenger and J. J. Pancrazio, *Biosens. Bioelectron.*, 2001, **16**, 535–542.
22. D. A. Stenger, G. W. Gross, E. W. Keefer, K. M. Shaffer, J. D. Andreadis, W. Ma and J. J. Pancrazio, *Trends Biotechnol.*, 2001, **19**, 304–309.
23. T. J. O'shaughnessy and J. J. Pancrazio, *Anal. Chem.*, 2007, **79**, 8838–8845.
24. N. V. Kulagina, C. M. Mikulski, S. Gray, W. Ma, G. J. Doucette, J. S. Ramsdell and J. J. Pancrazio, *Environ. Sci. Tech.*, 2006, **40**, 578–583.
25. T. J. O'Shaughnessy, J. L. Liu and W. Ma, *Biosens. Bioelectron.*, 2009, **24**, 2365–2370.
26. H. J. Lin, P. T. Charles, J. D. Andreadis, A. M. Churilla, D. A. Stenger and J. J. Pancrazio, *Anal. Chim. Acta*, 2002, **457**, 97–108.
27. K. M. Shaffer, H. J. Lin, D. Maric, J. J. Pancrazio, D. A. Stenger, J. L. Barker and W. Ma, *Biosens. Bioelectron.*, 2001, **16**, 481–489.
28. D. N. Stratis-Cullum, S. McMasters, L. J. Sooter and P. M. Pellegrino, in *Chemical and Biological Sensing VIII*, ed. A. W. Fountain, Proceedings of

the Society of Photo-optical Instrumentation Engineers (SPIE), 2007, vol. 6554, pp. Z5540–Z5540.
29. S. Song, L. Wang, J. F. Li, C. Fan and J. Zhao, *TrAC: Trends in Analytical Chemistry*, 2008, **27**, 108–117.
30. S. Tombelli, M. Minunni and M. Mascini, *Biosens. Bioelectron.*, 2005, **20**, 2424–2434.
31. S. Tombelli, M. Minunni and M. Mascini, *Biomol. Eng.*, 2007, **24**, 191–200.
32. S. A. Piletsky and A. P. F. Turner, *Imprinted Polymers and Their Application in Optical Sensors*, Elsevier, Amsterdam, 2008.
33. S. A. Trammell, M. Zeinali, B. J. Melde, P. T. Charles, F. L. Velez, M. A. Dinderman, A. Kusterbeck and M. A. Markowitz, *Anal. Chem.*, 2008, **80**, 4627–4633.
34. B. J. Melde, B. J. Johnson and P. T. Charles, *Sensors*, 2008, **8**, 5202–5228.
35. K. E. Sapsford, C. M. Soto, A. S. Blum, A. Chatterji, T. W. Lin, J. E. Johnson, F. S. Ligler and B. R. Ratna, *Biosens. Bioelectron.*, 2006, **21**, 1668–1673.
36. M. Liebau, A. Hildebrandt, J. Hartmann, U. Rothe, J. Muthing and R. Neubert, in *Chemical Microsensors and Applications*, ed. S. Buttgenbach, Proceedings of the Society of Photo-optical Instrumentation Engineers (SPIE), 1998, vol. 3539, pp. 188–195.
37. W. Knoll, K. Bender, R. Forch, C. Frank, H. Gotz, C. Heibel, T. Jenkins, U. Jonas, A. Kibrom, R. Kugler, C. Naumann, R. Naumann, A. Reisinger, J. Ruhe, S. Schiller and E. K. Sinner, in *Polymer Membranes/Biomembranes*, Advances in Polymer Science Series, 2010, vol. 224, pp. 87–111.
38. B. K. Lee, H. Y. Lee, P. Kim, K. Y. Suh and T. Kawai, *Lab on a Chip*, 2009, **9**, 132–139.
39. A. Hirano-Iwata, M. Niwano and M. Sugawara, *TrAC: Trends in Analytical Chemistry*, 2008, **27**, 512–520.
40. A. Janshoff and C. Steinem, *Anal. Bioanal. Chem.*, 2006, **385**, 433–451.
41. H. T. Tien and A. L. Ottova, *J. Membr. Sci.*, 2001, **189**, 83–117.
42. I. Koper, *Molecular BioSystems*, 2007, **3**, 651–657.
43. K. Masunaga, S. Michiwaki, R. Izumi, P. Ivarsson, F. Björefors, I. Lundström, K. Hayashi and K. Toko, *Sens. Actuator, B Chem.*, 2008, **130**, 330–337.
44. H. Tanaka, M. Hanasaki, T. Isojima, H. Takeuchi, T. Shiroya and H. Kawaguchi, *Colloids and Surfaces B: Biointerfaces*, 2009, **70**, 259–265.
45. W. Knoll, A. Kasry, F. Yu, Y. Wang, A. Brunsen and J. Dostalek, *J. Nonlinear Opt. Phys. Mater.*, 2008, **17**, 121–129.
46. S. Brahim, D. Narinesingh and A. Guiseppi-Elie, *Biosens. Bioelectron.*, 2002, **17**, 973–981.
47. D. N. Kim, W. Lee and W. G. Koh, *J. Chem. Tech. Biotechnol.*, 2009, **84**, 279–284.
48. L. Y. Wu, B. M. Ross, S. Hong and L. P. Lee, *Small*, 2010, **6**, 503–507.
49. V. A. Petrenko and I. B. Sorokulova, *J. Microbiol. Methods.*, 2004, **58**, 147–168.
50. C. Tamerler and M. Sarikaya, *Acta Biomaterialia*, 2007, **3**, 289–299.

CHAPTER 7

# Sensing and Transduction

## 7.1 Introduction

Perhaps one of the earliest sectors of biotechnology and bio-inspired approaches to be exploited is that of sensors. The ability to sense the environment, measure a change or difference and translate that into a recognizable signal is highly evolved in Nature and biology produces a myriad of natural components and systems where sensing and transduction are integrated. However, this integration is a more difficult challenge for those attempting to exploit bio-inspired systems. From the defence and security perspective, the capability to sense and detect a threat is recognized as essential to soldiers on a battlefield or in other mission spaces.[1] Effective sensors would allow rapid response, provide improved situational awareness and surveillance and assist in detection, targeting or tracking.

Because of the potential of biosensing, the development of inexpensive sensor elements and transduction methods remains a major area of interest and investment both at the national level and in private industry. Funding communities, discussed in Chapter 2, continue to have a high interest in supporting low-cost and deployable man-made devices, with research and development focused on the highly desirable characteristics of selectivity, sensitivity, reproducibility and rapid response times that could be provided by bio-based sensors. The European Union Framework directives call specifically for new technologies in biomonitoring, with hundreds of funded projects on biosensors[2] and the US agencies DARPA and DTRA have considerable investments in sensor development for chemical and biothreat detection. Despite these efforts, there has been limited success to date in fielding systems with the specifications and reliability required for military use. A truly deployable SASS type system

Bio-inspired Materials and Sensing Systems
By Peter Biggins, John Hiltz and Anne Kusterbeck

Published by the Royal Society of Chemistry, www.rsc.org

would require features that include robustness, smart functions, low power requirements, autonomy and multi-analyte detection. Though analytical chemists have developed a wide array of instruments and techniques that can be used to indicate the presence (and even the concentration) of molecules at trace levels, many of these are expensive, are not suitable for miniaturization and are limited to laboratory environments. Many of them also require extensive sample handling by skilled operators.

Similar to the components discussed in previous chapters, there are many excellent review articles describing recent advances in sensing and transduction in detail.[3,4] Provided herein is an overview of the most common methods and those that are emerging for applications in bio-inspired sensing.

## 7.2 Transduction Defined

At its simplest, detection of any analyte of interest or target depends on a number of factors, including interaction of receptors or reactive surfaces (discussed in Chapter 6) with the target molecules at sufficient levels to produce a response. Transduction is the conversion of the receptor binding into a recognizable, specific signal. Thus, for a signalling event to take place and be recognized, there must first be a stimulus. In biology, the signal transduction occurs when a mechanical or chemical stimulus to a cell is translated into a specific cellular response. Signal transduction starts with a signal to a receptor and ends with a change in behaviour. The list of external stimuli used to elicit a response in Nature includes temperature, light, mechanical force, gradient of magnetic field, electrical potential and chemical composition of the environment.

In a biological system, recognition events and transduction also represent a cascade of reactions, with multiple interconnects, feedback loops and response modulation. As seen in Figure 7.1, even a relatively simple event, such as the response to odour, though highly ordered and specific, has amazing complexity and involves many receptors, intermediates and interconnected events. In contrast, most fabricated sensors developed to date rely on a single recognition event in which a given stimulus provides only a linear response. Figure 7.2 provides a view of two of the general approaches used when building sensors in which the response is either a result of a specific binding by a receptor at the molecular level or an observation of a change in behaviour, as in cell-based sensing.

## 7.3 Select Examples of Sensing and Transduction Approaches

The many transduction mechanisms available for sensing and reporting are as varied as the number of targets to be detected. They can be roughly divided into optical, magnetic, electrochemical, mass-sensitive and mechanical, with a number of subsets under these formats, as shown in Figure 7.3.

For a given sensing scenario, there are numerous choices to be made when matching the receptor to a transduction method and the detection formats

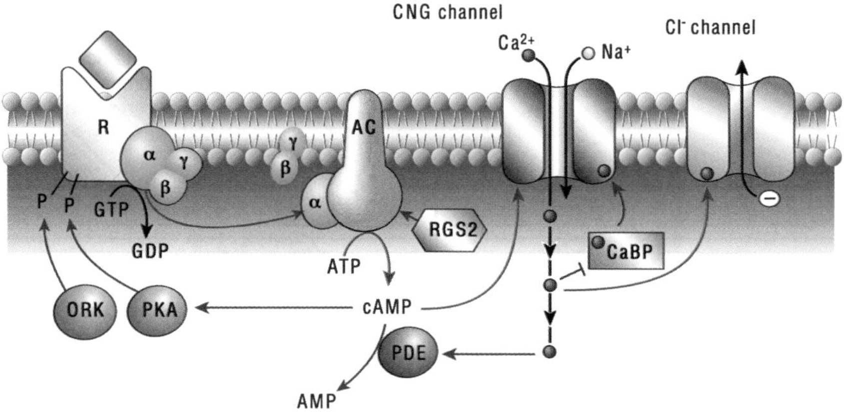
CNG channel
Ca2+
Na+
Cl- channel
R
α
γ
β
γ
β
AC
α
P
P
GTP
GDP
RGS2
CaBP
ATP
ORK
PKA
cAMP
PDE
AMP

**Figure 7.1** Example of the complexity of signal transduction in mammalian olfaction. Within the compact cilia of the Olfactory Sensory Neurons, a cascade of enzymatic activity transduces the binding of an odorant molecule to a receptor into an electrical signal that can be transmitted to the brain. This is a classic transduction pathway in which all of the proteins involved have been identified, cloned, expressed and characterized. AC: adenylyl cyclase; CNG channel: cyclic nucleotide-gated channel; PDE: phosphodiesterase; PKA: protein kinase A; ORK: olfactory receptor kinase; RGS: regulator of G proteins (but here acts on the AC); CaBP: calmodulin-binding protein. Green arrows indicate stimulatory pathways; red indicates inhibitory (feedback). (From A. Firestein, *Nature*, 2001, **413**, 211–218.)

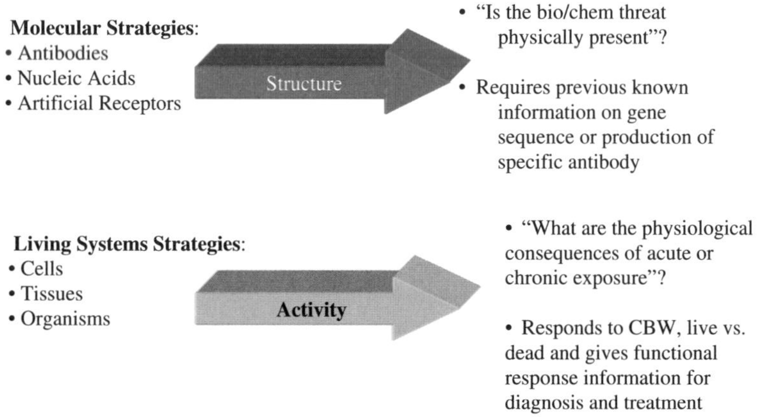
Molecular Strategies:
• Antibodies
• Nucleic Acids
• Artificial Receptors
Structure
• "Is the bio/chem threat physically present"?
• Requires previous known information on gene sequence or production of specific antibody
Living Systems Strategies:
• Cells
• Tissues
• Organisms
Activity
• "What are the physiological consequences of acute or chronic exposure"?
• Responds to CBW, live vs. dead and gives functional response information for diagnosis and treatment

**Figure 7.2** Comparison of a simple manmade sensor to a natural signal transduction scheme.

should be chosen in the context of the application. Optical methods may be most effective for biomolecules, while, in contrast, spectroscopic or electrochemical methods are preferred for chemical monitoring since they report based on a physical property of the target. Table 7.1 illustrates only a sampling of the broad

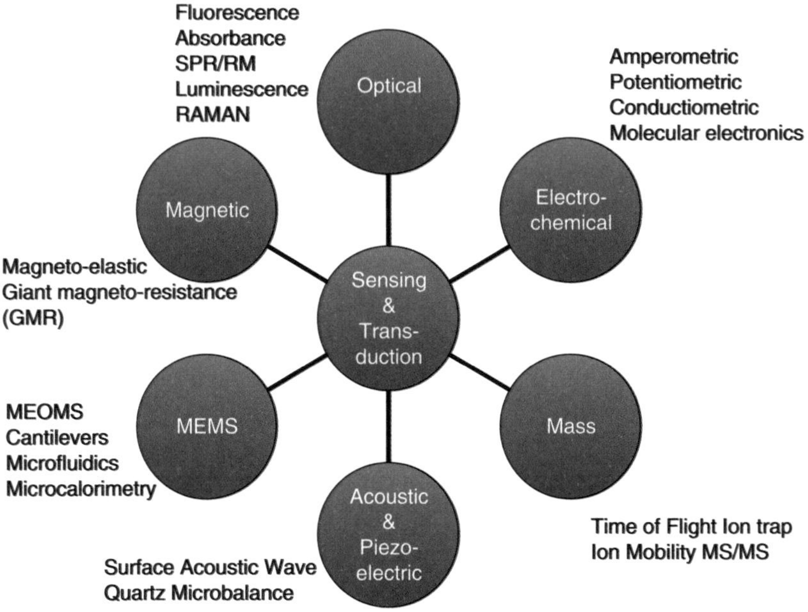
Fluorescence
Absorbance
SPR/RM
Luminescence
RAMAN
Optical
Amperometric
Potentiometric
Conductiometric
Molecular electronics
Magnetic
Electro-chemical
Magneto-elastic
Giant magneto-resistance
(GMR)
Sensing & Trans-duction
MEOMS
Cantilevers
Microfluidics
Microcalorimetry
MEMS
Mass
Acoustic & Piezo-electric
Time of Flight Ion trap
Ion Mobility MS/MS
Surface Acoustic Wave
Quartz Microbalance

**Figure 7.3** Detailed analysis of the variety and availability of sensing and transduction platforms.

array of analyte/transduction combinations that can be used in a variety of applications. The most commonly used methods that have potential applications in a conceptual SASS are discussed below.

### 7.3.1 Optical

Optical sensing and transduction methods comprise the largest majority of fielded systems due to the ready availability of low-cost lasers and photonic components. As shown in Figure 7.3, fluorescence, surface plasmon resonance (SPR) and colorimetric assays all fall within this category and the antibody-based sensors commonly used to detect biothreat agents rely most frequently on optical transduction methods to measure binding events. Figure 7.4 illustrates how the various techniques would work with antibodies immobilized on a substrate. In all cases, once the antibody binds the target, a change occurs in the optical output of the sensor that can be detected and quantified. The targets that have been detected range from single molecules to cellular organisms such as bacteria. For example, Kulinga *et al.*[5,6] have incorporated the natural antimicrobial peptides (AMPs) discussed previously into an array biosensor with fluorescent reporters. This biosensor can detect such biothreat agents as botulinum neurotoxoids A, B and E using AMPs as recognition elements.[5,6] Ligler *et al.*[7,8] have also been successful in harnessing fluorescence and

**Table 7.1** Examples of possible combinations of sensor formats, methods of transduction and applications.

| *Recognition element* | *Analytes* | *Signal transduction* | *Applications* |
|---|---|---|---|
| Antibodies | Explosives | SPR | Environmental monitoring |
| Peptides | Glucose | Optical | Trace detection |
| Nucleic acids | Metals | Electrochemical | Treaty verification |
| Mammalian cells | Drugs | Amperometric | Remote sensing |
| Bacteria | CBW agents | AFM | Medical diagnostics |
| Enzymes | NO | Magnetic | Border Security |
| Molecularly imprinted Polymers | Pesticides | Chemiluminescence | Detection of IED |

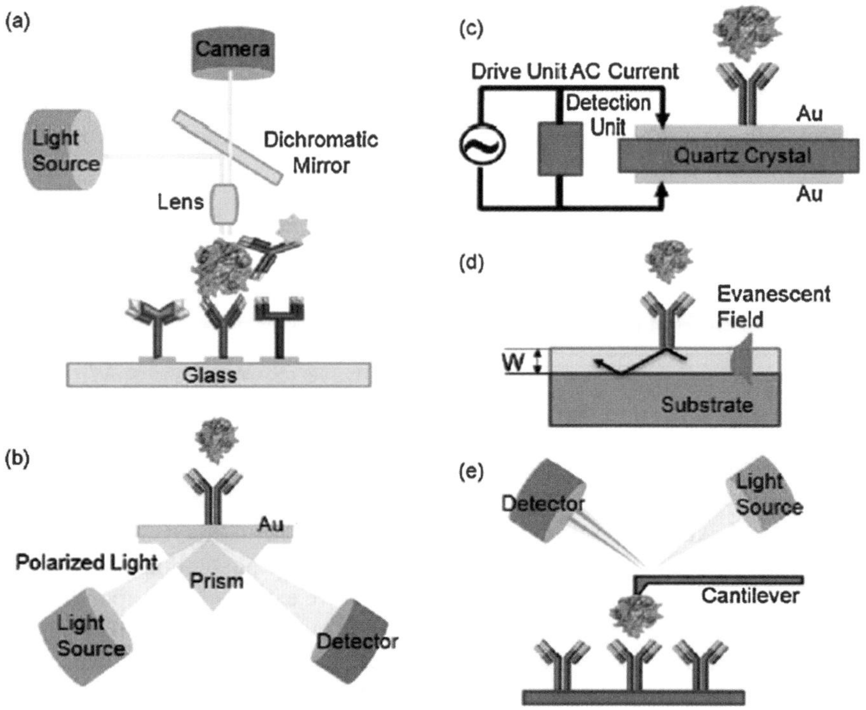
(a)
Camera
Light Source
Dichromatic Mirror
Lens
Glass
(b)
Au
Polarized Light
Prism
Light Source
Detector
(c)
Drive Unit AC Current
Detection Unit
Au
Quartz Crystal
Au
(d)
Evanescent Field
W
Substrate
(e)
Detector
Light Source
Cantilever

**Figure 7.4** Schematic overview of surface-based optical transduction platforms for pathogen detection. (a) fluorescence microscopy, (b) surface plasmon resonance (SPR), (c) quartz crystal microbalance (QCM), (d) optical waveguide sensors and (e) atomic force microscopy (AFM). (From A. W. Wark, J. Lee, S. Kim, S. N. Faisal and H. J. Lee, *J. Ind. Eng. Chem.*, **16**, 169. Copyright Elsevier.)

evanescent wave technologies to develop optical biosensors for field use in the rapid detection of chemical and biological targets.[7–9]

SPR optical methods are gaining ground based on the fact that, unlike the sensors above that use fluorescent dyes, no label is required and the kinetics of binding can be monitored. In this case, the binding of antibodies to a target produces an immediate change in the refractive index that can be quantified in real time, allowing kinetic studies of molecular interactions. As with the other optical methods, SPR has been demonstrated to detect many small molecules of interest in biosecurity (see the review by Shankaran *et al.*[10]).

### 7.3.2 Mass-based and Spectroscopic Methods

Though less amenable to portability and field use, the mass-based systems have gained popularity as they have become smaller. Because mass is a fundamental property of any material to be detected, this type of transduction is widely applicable for the detection of chemicals, including explosives and chemical agents. The major limitation of this technology is the lack of specificity and potential for false negatives since unknowns may not be part of the library used to identify targets. Hayden and colleagues have overcome these limitations using synthetic polymer receptors (the MIPs of Chapter 6) on microbalances to detect cells, viruses and enzymes.[11,12] Using a microfabricated ion mobility spectrometer, other groups have demonstrated the ability to distinguish Bacillus spores, chemical agents and bacteria.[13]

### 7.3.3 Piezoelectric

The primary technologies in this category are the well-studied quartz crystal microbalances (QCM) and surface acoustic wave (SAW) devices. With miniaturized devices and microfabrication capabilities, recent work has focused on sensitive microcantilevers as employed in atomic force microscopy (AFM). In all cases, a selective surface is incorporated into the sensor that detects the mass of the analyte upon binding. The versatility of this transduction is demonstrated by Dickert and colleagues who have successfully used QCM analysis for chemicals, viruses, enzymes and cells by incorporating biomimetic surfaces composed of artificial antibodies (MIPs) and have extended the capabilities to multiplexed detection on microfluidic chips.[14–19] Another example of cantilever application is the detection scheme developed by Shekhawat *et al.*,[20] which is a label- and optics-free detection paradigm and is based on metal-oxide semiconductor field-effect transistor (MOSFET)-embedded microcantilevers. Extremely small microcantilever bending, as small as 5 nm, occurs when there is a receptor-target binding event, leading to change in the MOSFET drain current. This group has been able to measure DNA hybridization and protein–protein binding with high sensitivity and they suggest that this may be a universal platform for the integration of nano-bio structures with microelectronics. Other groups have demonstrated detection of biothreat agents using a parallel acoustic/piezoelectric immunochip to detect multiple analytes.[21] When considering the SASS concept, this form of sensing and transduction may be amenable to the development of directionality or active touch in robots, as in the studies of Prescott *et al.*[22] designed to mimic the whiskering actions of a rat to sense the environment.

### 7.3.4 Electrochemical

Due to the rapid response time, direct measurement, low cost and connectivity with sensor output, electrochemical detection represents an extremely attractive transduction method. These types of sensors, second only to optical methods in popularity, measure electrochemical changes that occur when chemicals interact with a sensing surface of the detecting electrode. Different groups have exploited various strategies to develop sensors for analytes that range from explosives[23] to DNA.[24] The DNA-based electrochemical sensors in particular take advantage of the nanoscale interactions between the molecules in solution, the recognition layer and a solid electrode surface.

### 7.3.5 Micro-electromechanical Systems (MEMS)

A step beyond the basic electrochemical transduction, MEMS represents a true integration of the biological receptor with the microfabricated electronics to produce a simple sensor or actuator (see Zougagh and Rios for a review[25]). The bio-electronic combination not only provides improved functionality but, as suggested by Wu and Payne, may also offer the potential to use biological material in the actual construction process at the nanoscale.

### 7.3.6 Magnetic

Magnetic sensing and transduction is accomplished using magnetic particles that have a receptor bound to the surface. By applying a magnetic field gradient, these functionalized nanoscale beads can then be aligned, separated from unlabeled material or transported to another location. As an example, using a chip-based sensor with microfabricated current wires and giant magneto resistance (GMR) sensors, Janssen *et al.*[26] have demonstrated the real-time on-chip detection and manipulation of superparamagnetic particles in solution as part of a biosensor that can give information on biological function. Other groups have increased assay sensitivity by first releasing specifically bound magnetic particles, then actively guiding the unbound beads to the sensing zone in a location that gives rise to a theoretical maximum signal.[27]

### 7.3.7 Emerging Transduction Technologies

Within the past 5–10 years, one of the single most significant advances in sensing and transduction has been the introduction of nanotechnology. By taking advantage of the special properties afforded at the nanoscale, new methods are emerging that will change the way future sensors and sensing systems are developed.

#### *7.3.7.1 Silicon Nanowires and Nanoparticles*

Conventional detection systems involve labelling the probe molecules, a large amount of target molecules to enable detection and elaborate signal transduction methods. Most of them also have to couple with optical detection equipments that are bulky and expensive.

Maki *et al.*[28] have developed a molecular signal transduction system that 'translates' biomolecular interactions into universal signal molecules which are subsequently sensed by a nanotransistor. Detection of nucleic acids and protein from *S. aureus* was demonstrated on a universal sensing surface of a nano-FET.

Another emerging nanowire sensor has been demonstrated by J. Liu *et al.*[29] Unlike classical sensor systems, these one-dimensional (1-D) and two-dimensional (2-D) structures such as nanowires, nanobelts and films are capable of detecting the molecular interactions in terms of significant change in their electrical properties, thus leading to ultra-high sensitivity and easy integration. Using these small-scale sensors, they were able to demonstrate that ZnO nanowires-based biosensors could detect IgG antibodies.

Somers *et al.*[30] have reported semiconductor nanocrystals (NCs) as optical imaging agents where the photophysical properties of the NCs are insensitive to species in their environment.[30] This group modified NCs at the surface with a conjugate, which in turn interacts with an external agent by direct physical binding or chemical means. For sensing and transduction to take place, an energy transfer occurs between the NC and the target, which is the energy transfer acceptor or donor. As they demonstrated, signalling can be obtained by directly detecting luminescence from the NC and/or the conjugate. Finally, they suggest that the use of NCs as gain materials in micro-lasing cavities (distributed feedback gratings and spherical resonators) offers promise for CB-sensitive NCs for high-gain sensing applications.

### 7.3.8 Microfabrication and Lab on a Chip Technologies

The promise provided by design and engineering of complete sensing and transduction on a single microfabrication platform, the lab on a chip (LOC) concept, has spurred a whole new area of research. The benefits of these LOC technologies include small-scale measurements, disposable and low-cost components, as well as precise engineering with all components embedded on the chip. For example, Garnier *et al.*[31] have demonstrated femtomol detection levels on a conjugated polymer-based DNA chip with real-time access. This multiplot DNA-biosensor on a plastic chip was fabricated using the controlled electropolymerization of a thin film of an electroactive polymer on eight gold microelectrodes deposited on the chip. K. Misiakos *et al.*[32] have developed a

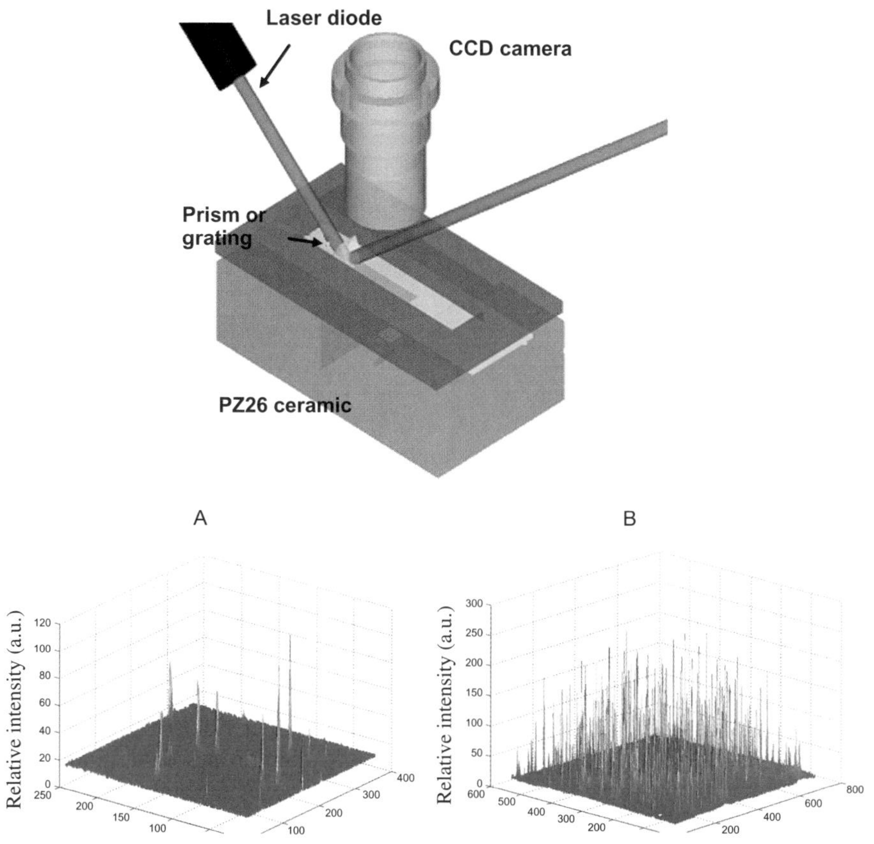
Laser diode
CCD camera
Prism or grating
PZ26 ceramic
A
B
Relative intensity (a.u.)
Relative intensity (a.u.)

**Figure 7.5** Example of Lab on a Chip (LOC) technology. Using integrated optics and microfluidics, assay detection thresholds were improved 200-fold. (courtesy of Martin McDonnell, Dstl.)

monolithic silicon optoelectronic transducer for protein sensing in which the biomolecular receptors are immobilized on microfluidic modules containing the optocouplers. Signal transduction is induced by the optical coupling deduction when the biomolecular binding event occurs. Shown in Figure 7.5 is an example of lab on a chip technology developed with integrated components, allowing 200-fold improvement in the detection limit.

## 7.5 Biomimetic and Bio-inspired Sensing Technologies

While multiple types of transduction mechanisms have been developed that can be applied to straightforward sensing of a single target, the real challenge in building the conceptual SASS outlined in Chapter 3 is how best to recreate the complexity, discrimination and responsiveness of biological systems to various stimuli as demonstrated by the five senses. Generally speaking, the senses can be divided into two classes: 1) the physical senses of vision, touch and hearing and 2) the chemical senses of smell and taste.[33] Although distinct in the ways in which these artificial systems are constructed, a common theme is a bio-inspired approach combining multiple arrays of sensing elements, pattern recognition and complex algorithms designed to reproduce a particular capability. Past efforts to fabricate realistic senses have met with limited success. Recently, however, as major advances have been made in networked arrays, miniaturization and computing power, researchers are beginning to achieve significant improvements. Most of these bio-inspired approaches for sensing involve extensive material development as well as engineering of the organic-inorganic interface. In the view of McConney *et al.*,[34] it is suggested that the most effective way to take advantage of biological ingenuity is to use a bidirectional approach, in which engineering learns from the way Nature has improved function through small incremental changes compared to totally new designs.

### 7.5.1 SMART Materials in Sensing and Transduction

Key to the design and fabrication of artificial sensing is a whole new class of materials referred to as SMART materials. These include predominantly the polymers discussed in Chapter 4 as well as new polymeric elastomers to sense flow, vibration or touch;[35] functional polymeric membranes that mimic cell membranes and report pathogens with a colour change;[36] combinations of high-affinity ligands, self-assembled monolayers, shape-specific receptor surfaces, mesoporous superstructures and microbeads for transduction to report CBRNE threats;[37] and hybrid silica/polypeptide membranes that exhibit molecular-level control of permeability in response to an external stimulus such as heat.[38]

### 7.5.2 Sensing Technologies

Beyond generalized sensing, there is also a substantial body of research focused on the development of a specific biological sense, *i.e.* the five human senses of touch, taste, sight, hearing and smell. Efforts to mimic olfaction (highlighted in Chapter 5) represent one example of progress towards the development of an electronic nose (e-nose), but there is also substantial research towards the realization of an e-tongue.[33,39–46] The various e-tongue formats consist primarily of sensor arrays for liquid sensing with the goal of discriminating among a mixture of chemical compounds. Other groups have engineered enhanced electronic noses or tongues using an artificial model based on the mammalian brain that incorporates an artificial neural network and have shown that performance is much improved over earlier systems.[47]

Similar to the work in e-tongues, researchers are using bio-inspired or biomimetic approaches to develop vision capabilities, primarily as a means to provide directionality and recognition for robots. In these systems, sensors have been developed that record light as a stimulus, just as the eye does, with signal response and processing modelled on that of the retina.[48]

Finally, tactile sensors to reproduce the sense of touch have also made significant progress.[49–56] The exquisite ability of biological systems to sense

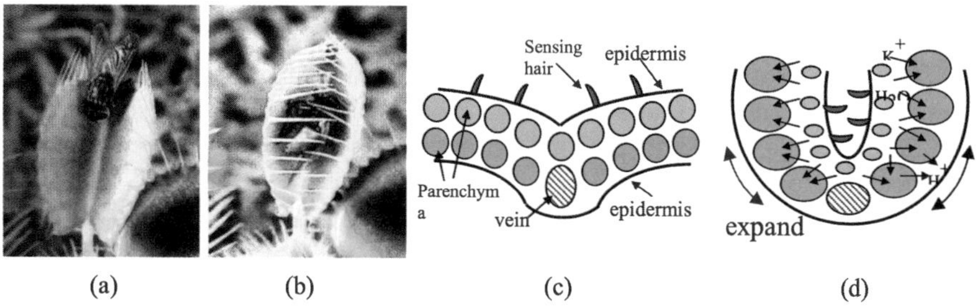
Sensing hair
epidermis
Parenchyma
vein
epidermis
expand
(a)
(b)
(c)
(d)

**Figure 7.6** Tactile sensing and transduction in Venus Fly Trap. Venus Fly Trap action in catching a flying insect, (a) insect touches the antenna located in the middle of leaf, (b) rapid closure of the leaves trap the insect, (c) cross section view before the leaf motion (d) after the leaf motion. (M. Taya, 2003, *Proc. SPIE*, **5051**, 54–65 with permission from SPIE.)

**Table 7.2** Selected bio-inspired/biomimetic approaches to sensing.

| *Sensing class* | *Sensing approach* | *Application* | *References* |
|---|---|---|---|
| Vision | Computer model of human system | Surveillance, vehicle steering system | Serre *et al.*[57,58] |
| | Vision chips and control algorithms modelled on insects | Autonomous flying robots | Zuffrey *et al.*[59] |
| Olfaction | Silicon chemFET array | Gas sensing | UK Universities |
| | Isolated olfactory cells on semiconductor | Odor sensing of chemicals | Q. Liu *et al.*;[60,61] Lee and Park;[62] Yoon *et al.*[63] |
| | Fluorescent optical fibres | Detection of toxic chemicals | Dickinson *et al.*[64] |
| | Yeast cells containing mammalian olfactory receptors | Dinitrotoluene (DNT), an explosive residue | Radhika *et al.*[65] |
| Tactile | Rigid core surrounded by fluid within elastomeric skin | Robotic devices | Wettels *et al.*[56] |
| | Artificial whiskers | Robotic touch | Prescott *et al.*[22] |
| Auditory | Electromechanical mimic of auditory hair cells | Highly sensitive force (sound) sensor | Song *et al.*[36] |
| Taste | Multichannel potentiometric electrodes with lipid-polymeric membranes | Electronic tongue system | Toko *et al.*[66,67] Iiyama *et al.*[68] |

touch is illustrated by the classic example of the Venus Fly Trap, a plant which feeds on insects that are detected using fine hairs on the surface, with subsequent actuation of the closing response (shown in Figure 7.6). Multiple groups are pursuing research to develop, for example, biomimetic whiskers for shape recognition[51] or MEMs-based multimodal tactile sensors made of polymer materials and metal thin film sensors to detect the hardness, thermal conductivity, temperature and surface contour of a contact object.[50]

The ability to reproduce the range and sensitivity of the senses has obvious benefits for the military and Table 7.2 provides a representative summary of these technologies and how they are being developed in areas of interest for the defence and security community.

## 7.6 Conclusion

Despite the ever-expanding list of sensing and transduction technologies, there remains a significant gap in the performance of these artificial sensors when compared to the inherent sensitivity, selectivity and rapid response of the biological systems. Limitations remain in realizing the full capabilities Nature has achieved due to difficulties that include, among others, systems integration, signal processing, receptor binding affinities, discrimination in complex mixtures and detection thresholds. Further progress in nanoscale fabrication and transduction mechanisms to mimic the five senses may enable the development of an SASS that meets operational requirements.

## References


1. R. E. Armstrong and J. B. Warner, in *Defense Horizons*, 2003, Gale Group, Farmington Hills, Michigan, pp. 1–8.
2. M. Farré, L. Kantiani, S. Pérez and D. Barcelo, *TrAC: Trends in Analytical Chemistry*, 2009, **28**, 170–185.
3. I. Tokarev, M. Motornov and S. Minko, *J. Mater. Chem.*, 2009, **19**, 6932–6948.
4. K. E. Sapsford, C. Bradburne, J. B. Detehanty and I. L. Medintz, *Mater. Today*, 2008, **11**, 38–49.
5. N. V. Kulagina, G. P. Anderson, F. S. Ligler, K. M. Shaffer and C. R. Taitt, *Sensors*, 2007, **7**, 2808–2824.
6. N. V. Kulagina, K. M. Shaffer, F. S. Ligler and C. R. Taitt, *Sens. Actuator, B Chem.*, 2007, **121**, 150–157.
7. F. S. Ligler, in *Optical Chemical Sensors*, ed. F. Baldini, A. N. Chester, J. Homola and S. Martellucci, 2006, vol. 224, pp. 437–455.
8. F. S. Ligler, *Anal. Chem.*, 2009, **81**, 519–526.
9. F. S. Ligler, K. E. Sapsford, J. P. Golden, L. C. Shriver-Lake, C. R. Taitt, M. A. Dyer, S. Barone and C. J. Myatt, *Anal. Sci.*, 2007, **23**, 5–10.

10. D. R. Shankaran, K. V. Gobi and N. Miura, *Sens. Actuator, B Chem.*, 2007, **121**, 158–177.
11. O. Hayden, R. Bindeus, C. Haderspock, K. J. Mann, B. Wirl and F. L. Dickert, *Sens. Actuator, B Chem.*, 2003, **91**, 316–319.
12. O. Hayden, P. A. Lieberzeit, D. Blaas and F. L. Dickert, *Adv. Funct. Mater.*, 2006, **16**, 1269–1278.
13. M. D. Krebs, B. Mansfield, P. Yip, S. J. Cohen, A. L. Sonenshein, B. A. Hitt and C. E. Davis, *Biomol. Eng.*, 2006, **23**, 119–127.
14. F. L. Dickert and O. Hayden, Molecularly Imprinted Polymers: Man-made Mimics of Antibodies and their Applications in Analytical Chemistry, in *Techniques and Instrumentation in Analytical Chemistry Series,* Elsevier, 2000, **vol. 23**, pp. 503–525.
15. F. L. Dickert, O. Hayden, P. Lieberzeit, C. Haderspoeck, R. Bindeus, C. Palfinger and B. Wirl, *Synthetic Met.*, 2003, **138**, 65–69.
16. F. L. Dickert, P. Lieberzeit, S. G. Miarecka, K. J. R. Mann, O. Hayden and C. Palfinger, *Biosens. Bioelectron.*, 2004, **20**, 1040–1044.
17. F. L. Dickert, P. Lieberzeit and M. Tortschanoff, *Sens. Actuator, B Chem.*, 2000, **65**, 186–189.
18. O. Hayden, R. Bindeus, C. Haderspöck, K.-J. R. Mann, B. Wirl and F. L. Dickert, *Sens. Actuator, B Chem.*, 2003, **91**, 316–319.
19. M. Jenik, A. Seifner, S. Krassnig, K. Seidler, P. A. Lieberzeit, F. L. Dickert and C. Jungbauer, *Biosens. Bioelectron.*, 2009, **25**, 9–14.
20. G. Shekhawat, S. H. Tark, V. P. Dravid and IEEE, *MOSFET-embedded microcantilevers: An all-electronic label- and optics-free signal transduction paradigm for bio-chem sensing*, 2006, vols 1–3, pp. 746–748, ISBN 978-1-4244-0375-2.
21. T. Alava, N. Berthet-Duroure, C. Ayela, E. Trévisiol, M. Pugnière, Y. Morel, P. Rameil and L. Nicu, *Sens. Actuator, B Chem.*, 2009, **138**, 532–538.
22. T. J. Prescott, M. J. Pearson, B. Mitchinson, J. C. W. Sullivan and A. G. Pipe, *IEEE Robot. Autom. Mag.*, 2009, **16**, 42–50.
23. S. Trammell, F. Velez, P. Charles and A. Kusterbeck, *Anal. Lett.*, 2008, **41**, 2634–2645.
24. T. G. Drummond, M. G. Hill and J. K. Barton, *Nat. Biotechnol.*, 2003, **21**, 1192–1199.
25. M. Zougagh and A. Rios, *Analyst*, 2009, **134**, 1274–1290.
26. X. J. A. Janssen, L. J. van Ijzendoorn and M. W. J. Prinse, *Biosens. Bioelectron.*, 2008, **23**, 833–838.
27. R. Wirix-Speetjens, G. Reekmans, R. De Palma, C. Liu, W. Laureyn and G. Borghs, *Sens. Actuator, B Chem.*, 2007, **128**, 1–4.
28. W. C. Maki, N. N. Mishra, S. K. Rastogi, E. Cameron, B. Filanoski, P. Winterrowd and G. K. Maki, *Sens. Actuator, B Chem.*, 2008, **133**, 547–554.
29. J. Liu, J. Goud, P. M. Raj, M. Iyer, Z. Wang, R. R. Tummala and IEEE, in *57th Electronic Components & Technology Conference, 2007 Proceedings*, 2007, pp. 1971–1976, ISBN 978-1-4244-0984-6.

30. R. C. Somers, M. G. Bawendi and D. G. Nocera, *Chem. Soc. Rev.*, 2007, **36**, 579–591.
31. F. Garnier, B. Bouabdallaoui, P. Srivastava, B. Mandrand and C. Chaix, *Sens. Actuator, B Chem.*, 2007, **123**, 13–20.
32. K. Misiakos, P. S. Petrou, S. E. Kakabakos, M. E. Vlahopoulou, A. Tserepi, E. Gogolides and H. H. Ruf, *Microelectronic Engineering*, 2006, **83**, 1605–1608.
33. P. Ciosek and W. Wroblewski, *Analyst*, 2007, **132**, 963–978.
34. M. E. McConney, K. D. Anderson, L. L. Brott, R. R. Naik and V. V. Tsukruk, *Adv. Funct. Mater.*, 2009, **19**, 2527–2544.
35. C. Liu, J. Engel, J. Chen, N. Chen, S. Pandya, Y. Yang, C. Tucker, S. Coombs, J. Humphrey, H. Bleckmann and IEEE, *IEEE International Conference of Nano/Micro Engineered and Molecular Systems*, Zhuhai, People's R. China, 2006.
36. J. Song, Q. Cheng, S. M. Zhu and R. C. Stevens, *Biomedical Microdevices*, 2002, **4**, 213–221.
37. M. Lind and B. Wright, *SPIE Conference on Nondestructive Detection and Measurement for Homeland Security*, San Diego, CA, 2004, pp. 1–7.
38. G. V. R. Rao, S. Balamurugan, D. E. Meyer, A. Chilkoti and G. P. Lopez, *Langmuir*, 2002, **18**, 1819–1824.
39. F. P. A. Cabral, B. B. Bergamo, C. A. R. Dantas, A. Riul and J. A. Giacometti, *Rev. Sci. Instrum.*, 2009, **80**.
40. B. E. Collins, A. T. Wright and E. V. Anslyn, in *Creative Chemical Sensor Systems*, 2007, vol. 277, pp. 181–218.
41. V. Koleshko, N. Khmurovich, Y. Varabei, E. Polynkova and IEEE, *Biosensor Intelligent System of Rapid Diagnosis*, 2009, pp. 26–30.
42. S. Litvinchuk, H. Tanaka, T. Miyatake, D. Pasini, T. Tanaka, G. Bollot, J. Mareda and S. Matile, *Nat. Mater.*, 2007, **6**, 576–580.
43. A. Mimendia, J. M. Gutiérrez, L. Leija, P. R. Hernández, L. Favari, R. Muñoz and M. del Valle, *Environ. Modelling Software*, 25, 1023–1030.
44. Y.-S. Sohn, A. Goodey, E. V. Anslyn, J. T. McDevitt, J. B. Shear and D. P. Neikirk, *Biosens. Bioelectron.*, 2005, **21**, 303–312.
45. H. S. Song, T. H. Kim, S. H. Lee, U.-K. Kim, S. Hong and T. H. Park, *J. Biosci. Bioeng.*, 2009, **108**, S152–S152.
46. P. Wang, Q. J. Liu, W. Zhang and H. Cal, *Sensor. Mater.*, 2007, **19**, 309–323.
47. G. Pioggia, M. Ferro, F. Di Francesco, A. Ahluwalia and D. De Rossi, *Bioinspiration & Biomimetics*, 2008, **3**, 016004.
48. H. Okuno, T. Yagi and IEEE, *SICE-ICASE International Joint Conference*, Busan, South Korea, 2006.
49. R. Bischoff and V. Graefe, in *Intelligent Autonomous Systems 9*, Eds. T. Arai, R. Pfeifer, T. Balch and H. Yokoi, IOS Press, Amsterdam, 2006, pp. 124–132.
50. J. Engel, J. Chen, Z. Fan and C. Liu, *Sensor. Actuator. A Phys.*, 2005, **117**, 50–61.
51. D. Kim and R. Möller, *Robot. Autonom. Syst.*, 2007, **55**, 229–243.

52. R. Maldonado-Lopez, F. Vidal-Verdu, G. Linan, E. Roca and A. Rodriguez-Vazquez, *Conference on Bioengineered and Bioinspired Systems II*, Seville, Spain, 2005.
53. R. Maldonado-Lopez, F. Vidal-Verdu, G. Linan, E. Roca and A. Rodriguez-Vazquez, *Analog Integrated Circuits and Signal Processing*, 2007, **53**, 97–108.
54. J. Scheibert, S. Leurent, A. Prevost and G. Debregeas, *Science*, 2009, **323**, 1503–1506.
55. J. Wang, H. Sato, C. Y. Xu and M. Taya, *J. Appl. Phys.*, 2009, **105**, 083515.
56. N. Wettels, V. J. Santos, R. S. Johansson and G. E. Loeb, *Adv. Robot.*, 2008, **22**, 829–849.
57. T. Serre, L. Wolf, S. Bileschi, M. Riesenhuber and T. Poggio, *IEEE Transactions on Pattern Analysis and Machine Intelligence*, 2007, **29**(3), 411–426.
58. T. Serre, G. Kreiman, M. Kouh, C. Cadieu, U. Knoblich and T. Poggio, *Progress in Brain Research*, 2007, **165**, 33–56.
59. J. C. Zufferey and D. Floreano, *IEEE Trans. Robotics*, 2006, **22**(1), 137–146.
60. Q. Liu, W. Ye, H. Yu, N. Hu, L. Du, P. Wang and M. Yang, *Sens. Actuators, B Chem.*, 2010, **146**, 527–533.
61. Q. Liu, J. Yu, L. Xiao, J. C. O. Tang, Y. Zhang, P. Wang and M. Yang, *Biosens. Bioelectron.*, 2009, **24**, 1305–1310.
62. S. H. Lee and T. H. Park, *Biotech. Biopro. Eng.*, 2010, **15**, 22–29.
63. H. Yoon, S. Lee, O. Kwon, H. Song, H. EH, T. Park and J. Jang, *Angew. Chem. Int.*, 2009, **48**, 2755–2578.
64. T. Dickinson, J. White, J. Kauer and D. Walt, *Nature*, **382**, 1996, pp. 697–700.
65. V. Radhika, T. Proikas-Cezanne, M. Jayaraman, D. Onesime, J. H. Ha and D. N. Dhanasekaran, *Nature Chemical Biology*, 2007, **3**, 325–330.
66. K. Toko, T. Matsuno, K. Tamafuji, K. Hayashi, H. Ikezaki, K. Sato, R. Toukubo and S. Kawarai, *Biosens. Bioelectron.*, 1994, **9**, 359–364.
67. K. Toko, *Sens. Actuator, B Chem.*, 2000, **64**, 205–215.
68. S. Iiyama, S. Ezaki and K. Toko, *Sens. Actuator, B Chem.*, 2009, **141**, 343–348.

CHAPTER 8
# *Energy and Power*

## 8.1 Introduction

In the defence and security context, the majority of fielded sensor systems require power whether it is from a mains supply, power generators or batteries. This requirement exists irrespective of the size and complexity of the sensor system. In fact, the supply of power and energy is a major issue within the defence sector as there is a growing awareness that alternative energy sources are required to replace our dependency on those sources that depend on diminishing fossil fuel resources. Whilst economics influence the rate and level of development of alternative power generation systems in the civil sector, the changes in economics of energy supply are starting to favour novel or alternative approaches to power generation. These include the use of wind, tidal and solar energy sources to produce power.

The conventional approach to the supply of continuous power requires the storage of a fuel and the conversion of that fuel into power. The everyday large-scale examples are coal-, oil- and gas-fired electricity generating stations and nuclear power stations feeding into fixed electrical distribution infrastructures.

The viability of remote autonomous sensors requires that a reliable source of power be available. The provision of power is crucial as it affects the length of time the sensor can be in service and the level of functionality that can be incorporated into a sensor. The issue of power and its re-supply is a potential weakness for remotely deployed systems. This is typified by the satellite and space probe systems where the lifetime of the power source dictates the lifetime of the system.

Within the military context, energy sources are crucial for operations and are usually dependent on the type of military activity being undertaken. For certain types of surveillance activity, there is a need to develop sensor systems that are

---

Bio-inspired Materials and Sensing Systems
By Peter Biggins, John Hiltz and Anne Kusterbeck

Published by the Royal Society of Chemistry, www.rsc.org

energy efficient. If they are deployed in remote locations for long periods of time they must either be self-sustaining or the power source must be replaced or replenished. This latter point may also be true for certain activities within a secure area. However, conventional security surveillance systems will usually have access to central power supplies, although this will depend on the state of the power distribution infrastructure.

The development of reliable, low-mass sources of power and energy is critical for deployed sensor systems. For extended periods of operation, the available energy is likely to be the limiting parameter, but if the system has to move about in its environment then peak power may also have to be considered. Battery technology has advanced significantly in recent years, but battery-powered systems still have relatively short lifetimes. In addition, the logistics burden imposed by battery usage is a growing problem for the military. The appetite for power is a major issue for the future soldier and is receiving ongoing attention.[1] The scope for dramatic improvements in battery systems is limited by the useable battery chemistries. This leads one to ask the following question; are there any alternatives suggested by natural systems?

Unmanned systems demand an even higher level of design and engineering for power since, by design, there is a requirement for a specific level of autonomy. As seen in Figure 8.1, Morgan *et al.*, have calculated the power

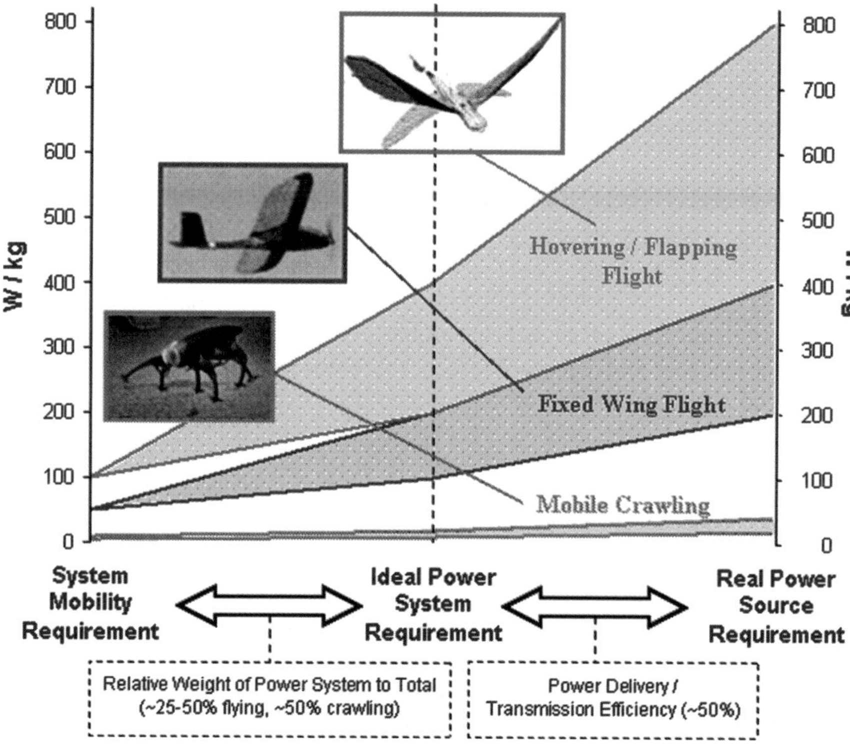
W / kg
800
700
600
500
400
300
200
100
0
Hovering / Flapping Flight
Fixed Wing Flight
Mobile Crawling
System Mobility Requirement
Ideal Power System Requirement
Real Power Source Requirement
Relative Weight of Power System to Total (~25-50% flying, ~50% crawling)
Power Delivery / Transmission Efficiency (~50%)

**Figure 8.1** Specific Power (Avg) required for Mobile Microsystems (< 10 W).[2]

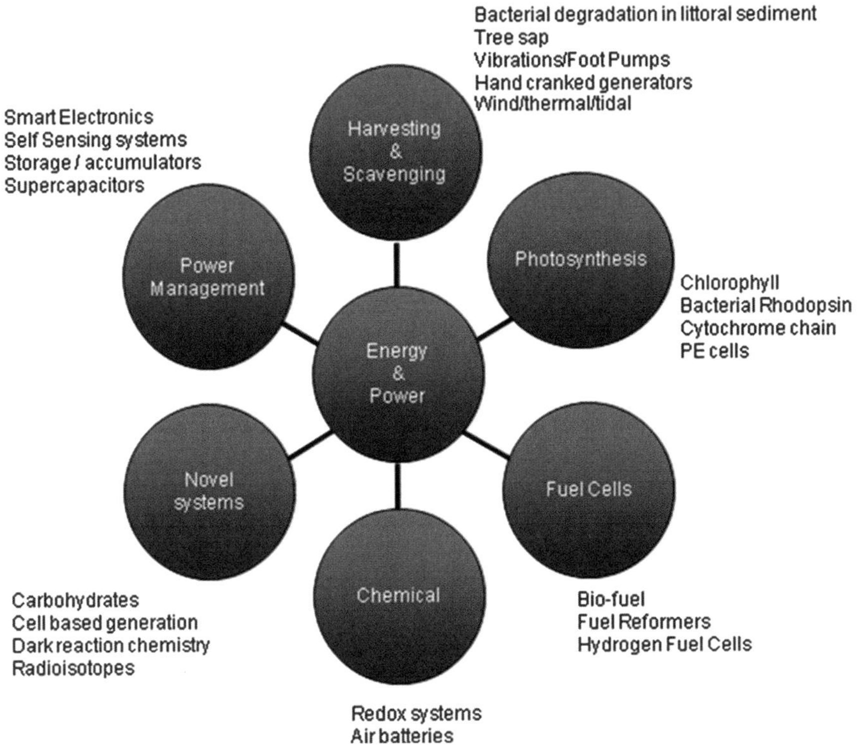
Bacterial degradation in littoral sediment
Tree sap
Vibrations/Foot Pumps
Hand cranked generators
Wind/thermal/tidal
Smart Electronics
Self Sensing systems
Storage / accumulators
Supercapacitors
Harvesting
&
Scavenging
Power
Management
Photosynthesis
Chlorophyll
Bacterial Rhodopsin
Cytochrome chain
PE cells
Energy
&
Power
Novel
systems
Fuel Cells
Chemical
Carbohydrates
Cell based generation
Dark reaction chemistry
Radioisotopes
Bio-fuel
Fuel Reformers
Hydrogen Fuel Cells
Redox systems
Air batteries

**Figure 8.2** Energy and power for a SASS.

required to operate unmanned systems of various weights, payloads and flight durations and have shown a clear relationship between operational requirements and energy demands.[2] The potential sources of power and approaches to generating power for an SASS, including fuel cells, energy harvesting, photosynthesis, chemicals and storage/accumulators, are shown in Figure 8.2. In addition to power and energy sources, the management of the energy and power production is also crucial to the operation of these systems.

## 8.2 Energy Sources

The sun is the primary energy source for sustaining life and biological systems are very good at using energy derived from the sun for survival. In this chapter, the discussion is focused on deriving energy and power from natural sources or on using bio-inspired approaches to generate energy and power. Several bio-inspired approaches to the production of energy and power, including photosynthesis, photovoltaic devices and microbial fuel cells, are examined.

The sources of energy for powering sensing platforms are varied. The methods and developing technologies to harvest these sources are described below.

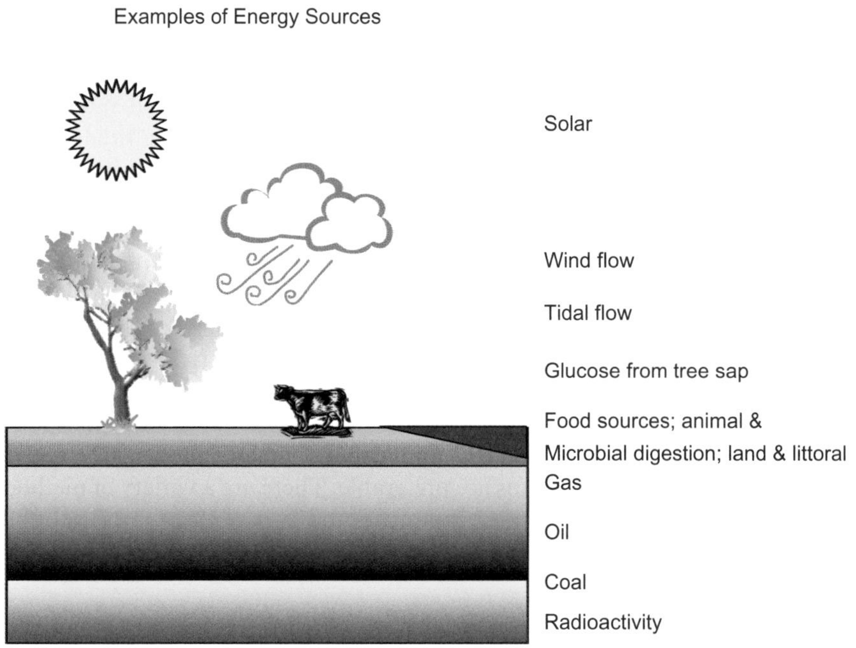
Examples of Energy Sources
Solar
Wind flow
Tidal flow
Glucose from tree sap
Food sources; animal &
Microbial digestion; land & littoral
Gas
Oil
Coal
Radioactivity

**Figure 8.3** Sources of energy.

### 8.2.1 Energy in a Natural System

Figure 8.3 depicts some of the principal sources of energy from Nature. Photosynthesis, which takes place in the tree and the carpet of grass where the cow is standing, converts the energy from sunlight into biomass. When the biological system dies, the biomass decays and over geological timeframes is eventually transformed into oil, gas and coal deposits. These fossil fuels represent stored energy, which is released through combustion or catalytic processes to provide power generation.

Technologies to harvest energy from the environment in a more efficient way are gaining wider usage as energy sources, such as the sun, tidal flow, river flow and wind, are being used to power systems. Energy harvesting is not a new concept and throughout our history there have been many examples of systems developed to use energy from natural sources, such as wind and water flow, to power devices. These systems range from windmills to hydroelectric generators and are usually large scale and static.

Observations of how seaweed moves have led one firm to develop a tidal energy conversion system, BioWAVE™,[3] which is based on the swaying motion of seaweed. This company has also developed a tidal energy conversion system, Biostream™, which mimics the shape and motion characteristics of sharks. It is reported that sharks are nearly ninety percent efficient in converting body movement into propulsive force. In a fixed-location generating system based on the shark, the propulsion mechanism is reversed such that energy in the passing tidal flow is used to drive the device motion against the resisting torque of the generator.

The use of solar cells for generating power has been elegantly demonstrated by the Qinetiq Zephyr programme. The Zephyr is a high-altitude, long-endurance, unmanned air vehicle (UAV) built out of ultra-light weight carbon fibre composite. The power for flight is provided by solar energy, which is collected by an array of United Solar ovonic amorphous silicon solar cells covering the aircraft's wings. The solar arrays are no thicker than sheets of paper and during daylight produce enough power to keep the vehicle aloft during the night by using rechargeable lithium sulfur batteries. The UAV has operated at over 58,000 feet for a period of 54 hours. It has considerable potential for applications such as earth observation and as a communications relay.[4]

For certain defence and security applications, small, unobtrusive and in some cases highly mobile systems are preferable. There are a variety of methods for harvesting energy from the environment, and two of these, artificial photosynthesis systems and microbial fuel cells, are discussed in the following section.

### 8.2.2 Solar Energy

The most direct approach to converting solar energy into electricity is through the use of solar cells. Some space probes and satellites rely on solar cells to provide power for on board equipment. Solar cells are usually silicon-based devices which convert the energy of sunlight directly into electricity using the photovoltaic effect. The terms solar cell and photovoltaic cell will be used interchangeably here, although the term *solar cell* is sometimes reserved for devices that capture energy from sunlight, while the term *photovoltaic cell* is used when the light source is unspecified.

The majority of biological organisms rely on sunlight, both directly and indirectly, for sustaining life. Therefore, our attempts to develop systems that derive power from the environment are examples of bio-inspired approaches.

Early organisms were probably heterotrophs, that is, they were able to synthesize their food in a non-oxygen (reducing) environment. During the early phase of the evolutionary process a vital source of energy for organisms was hydrogen. The early organisms that evolved in the conditions that prevailed were able to fix hydrogen. However, when the atmosphere changed from an oxygen-poor to an oxygen-rich atmosphere, the evolutionary process moved in the direction of photosynthetic systems (*photoautotrophs* are able to synthesize food directly from carbon dioxide using energy from light). It followed that those organisms capable of hydrogen fixation lost their evolutionary relevance. This resulted in organisms that were unable to develop oxygen-resistant hydrogenases. Whilst hydrogenases still exist in bacteria, such as cyan-bacteria and unicellular algae, the hydrogenase genes only switch on under anaerobic conditions.

### 8.2.3 Photosynthesis

One approach to developing a self-sustaining SASS system could be based on the principles of the photosynthetic pathway. Photosynthesis, both directly and indirectly, is crucial for the majority of life forms on the planet. It not only provides nourishment but also is responsible for maintaining the level of oxygen required to sustain life. It is the process by which plants, algae and some bacteria use solar energy to convert carbon dioxide and water into organic compounds, such as carbohydrates, oils and cellulose. The photosynthesis reaction is defined as:

$$6\,CO_2 + 12\,H_2O + \text{photons} >>>> C_6\,H_{12}O_6 + 6\,O_2 + 6\,H_2O$$

Sunlight is absorbed by proteins that contain chlorophyll, which are found in organelles called chloroplasts. Some of the light energy gathered by the chlorophyll is stored in the form of adenosine triphosphate (ATP).

Photosynthesis occurs in two stages. The first stage involves the light-dependent reactions in which the light energy is used to make ATP from adenosine diphosphate (ADP) and the reduced form of nicotinamide adenine dinucleotide phosphate (NADPH). This process has an efficiency of almost nine percent.[5]

The overall reaction is:

$$\begin{aligned} &2\,H_2O + 2\,NADP^+ + 3\,ADP + 3\,P_i + \text{light} \\ &\rightarrow 2\,NADPH + 2\,H^+ + 3\,ATP + O_2 \end{aligned}$$

The second stage involves the light-independent reactions, where $CO_2$ is taken from the atmosphere and used to produce sucrose and starch in a process described by the Calvin–Benson cycle.[6]

### 8.2.4 Artificial Photosynthesis

The photosynthetic processes that convert sunlight into energy can be replicated in artificial systems. The artificial systems use sunlight to split water into hydrogen and oxygen. The hydrogen produced by this process can be used to generate power in a fuel cell. Two approaches are currently used. These are photo electrochemical and dye-sensitized solar cells.

#### *8.2.4.1 Photochemical*

Using water as a source of electrons and sunlight as a source of energy, engineered natural systems and biomimetic (bio-inspired) model systems can be designed to allow water splitting based hydrogen production (bio-photolytic hydrogen production).[7] A photosynthetic solar cell, which resembles a Gratzel cell, mimics photosynthesis and has been used to produce hydrogen directly from water. Ruthenium-based dye molecules are attached to a catalyst particle so that when sunlight hits the dye, electrons are liberated and directed straight into the catalyst where they split water molecules into oxygen and hydrogen approximately 1000 times faster than other dye-based cells.[8]

Titanium disilicide acts as a photo catalyst, storage medium and separator for hydrogen produced from water. This inexpensive, easily obtained semiconductor absorbs light over a wide range of the solar spectrum. As a catalyst, it splits water into hydrogen and oxygen with a higher efficiency than other semiconductor systems that use visible light. Although its hydrogen storage capacity is less than that observed for usual storage materials, lower temperatures are needed for hydrogen release. For oxygen release, temperatures over 100 °C and darkness are needed.[9]

#### *8.2.4.2 Dye-sensitized Solar Cells*

New solar cells do not need direct sunlight to operate and can be made using a range of dyes. The cells are made from titanium dioxide which is abundant, non-toxic[9] and environmentally friendly. The dyes can be impregnated into roofing materials, window glass and even clothing and used to produce power. Novel dyes, including those based on porphyrins and haemoglobin,[10] are being tested in the prototype solar cells. The efficiency of the dyes at converting sunlight into electricity allows the cells to operate in low light conditions.

Solar panels have been made from organic materials which have efficiencies of four percent. For instance, they have been made with the pigments of blueberries; the idea is that organic semiconductors might significantly reduce the cost of solar panels.[11]

Solar cells can be derived from sea sponges, like the orange puffball. It synthesizes pristine layers of silica without the benefit of high temperatures or low pressures and at nearly neutral pH using an enzyme called silicatein. This enzyme catalyses the formation of silica structures from silicic acid found in seawater. Using this enzyme researchers have now developed an analogous low-energy process that produces structured layers of zinc oxide using aqueous zinc nitrate and ammonia. They have built crystalline layers of zinc oxide 100 to 300 nm thick on glass substrates.[12] Zinc oxide is widely used as a solar cell semiconductor.

### 8.2.5 Fuel Cells

The principal of a fuel cell was first described by C. F. Schobien in 1839 and demonstrated a year later by W. R. Grove.[13] However, it wasn't until the 1950s that the development of fuel cells started in earnest. This was the result of the recognition that fuel cells had many potential applications as power sources in remote locations. Typical applications included their use as power sources on board spacecraft and for remote weather stations. The advantage of a fuel cell system is that it can be compact and lightweight, and has no major moving parts. There are many different types of fuel cells and a considerable amount of research is being undertaken to develop more efficient fuel cells.[14,15] From a bio-inspired perspective, microbial fuel cells which are still in the research phase have the potential to power a SASS-type sensing system if they can harvest a source of fuel from the environment.

#### *8.2.5.1 Microbial Fuel Cells*

Microbial Fuel Cells (MFCs) are bio-electrochemical transducers that convert biochemical energy to electrical energy. These cells are based on nineteenth-century technology which demonstrated that electricity could be derived from bacteria.[16,17]

The principle of operation of an MFC is shown in Figure 8.4(a). Bacteria act as a catalyst to generate energy from the substrate at the MFC anode. Oxygen ($O_2$) acts as an oxidizing agent to take up the electrons and protons to produce water ($H_2O$) at the cathode. This closes the circuit and keeps the system balanced.

Comparison of the properties of a MFC and a standard AA alkaline battery indicates that a MFC cannot match the voltage, capacity, energy or energy density of an AA alkaline cell. However, the durability (lifetime) of the MFC can be considerably greater than the AA alkaline cell. The potential for the MFC to produce power for extended periods of time, provided the bacteria can be sustained, makes it an attractive option for remote sensing systems.

An example of research directed at developing autonomous robotic systems that generate their own power using $O_2$ cathodes open to the air MFCs (Figure 8.4(b),(c)) has been demonstrated by the Bristol Robotic Laboratories (BRL) by their EcoBot series.[19–21] The researchers have developed a robotic system called EcoBot II designed to power itself solely by converting unrefined insect biomass using an on-board MFC with oxygen cathodes. The system has demonstrated phototaxis and was able to transmit information on the thermal gradient as it powered the robotic system towards a light source. This concept has been further developed using the anaerobic bacteria *Geobacter sulphurreducens* in the BRL EcoBot III system. The same researchers have also investigated the possibility of exploiting MFCs with aqueous oxygen cathodes as artificial gills for autonomous robots.[22] The BRL EcoBot series demonstrates energetic autonomy by converting natural raw biological derived substrates such as flies, carrots or apples into a source of power. These systems require an artificial digestion system and concomitant artificial metabolism that can be achieved through the use of a MFC. In Figure 8.5(a),(b), the mean current output for different substrates, both refined and unrefined, is given.[23]

Power generation and storage techniques have been developed for use in marine environments based on the oxidation of sediment organic matter by oxygen from the overlying water. Although concentrations depend on location,

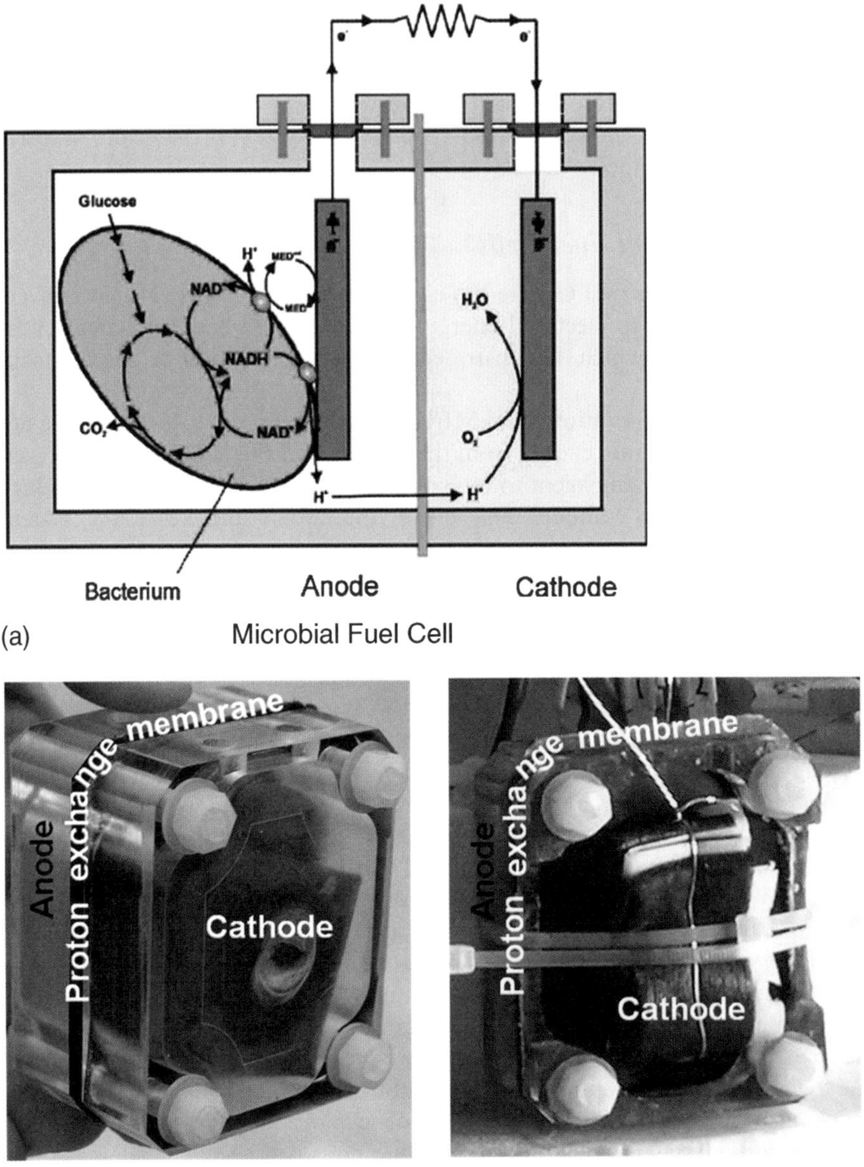
Glucose
NADH
Bacterium
Anode
Cathode
(a)
Microbial Fuel Cell
Proton exchange membrane
Anode
Cathode
Proton exchange membrane
Anode
Cathode
(b)

**Figure 8.4** a) A general layout of a Microbial Fuel Cell (MFC) shows the anodic compartment in which the bacteria brings about the oxidative conversions while in the cathodic compartment chemical and microbial reductive processes occur (after Rabaey and Verstraete, 2005[18]). b) Prototype MFCs showing the cathode open to the atmosphere (courtesy of I. Ieropoulos, J. Greenman and C. Methuish, BRL).[20] c) Schematic of the open to air cathode microbial fuel cell (courtesy of I. Ieropoulis, J. Greenman and C. Melthuish, BRL).[20]

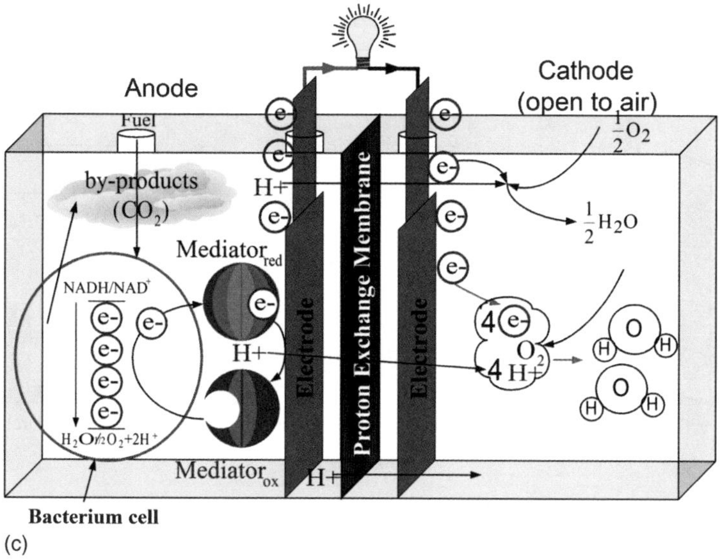
Anode
Fuel
by-products
(CO2)
Mediator red
NADH/NAD+
e-
H+
H2O/½O2+2H+
Mediator ox
Bacterium cell
Electrode
Proton Exchange Membrane
Electrode
Cathode
(open to air)
1/2 O2
1/2 H2O
4 e-
O2
4 H+
O
H
(c)

**Figure 8.4** Continued

marine sediment contains about two percent organic matter, which provides a significant energy source. Energy harvesting from estuarine and marine sediments to power unattended sensing platforms in rivers or the ocean has been demonstrated by the US Naval Research Laboratory (NRL). The Benthic Underwater Generator (BUG),[24] shown in Figure 8.6, has been designed to use underwater microbes to provide power and extend the lifetime of autonomous marine instrumentation.

The production of energy from a variety of sources such as tree sap and mammalian fluids has also been investigated.[25] Energy might also be scavenged from other sources such as sunlight, heat and vibration.[26,27] Another approach is to use an understanding of electric eel cells to develop power systems. Electric eels channel the output of thousands of specialized cells called electrocytes to generate electric potentials of up to 600 V. A complex numerical model of the conversion of ion concentrations to electrical impulses showed that an artificial cell could generate forty percent more energy in a single pulse than a natural electrocyte.[28]

In Figure 8.7, power production for MFCs is shown over time on the basis of published results. In less than a decade, power production by MFCs has increased by several orders of magnitude. Power production continues to be limited by systems that have the cathode immersed in water [aqueous cathodes (red triangles) and sediment MFCs (green diamonds)]. Substantial power production has been possible by using air-cathode designs in which the cathode is exposed to air on one side and water on the other side (blue squares). In general, wastewaters have produced less power than systems using pure

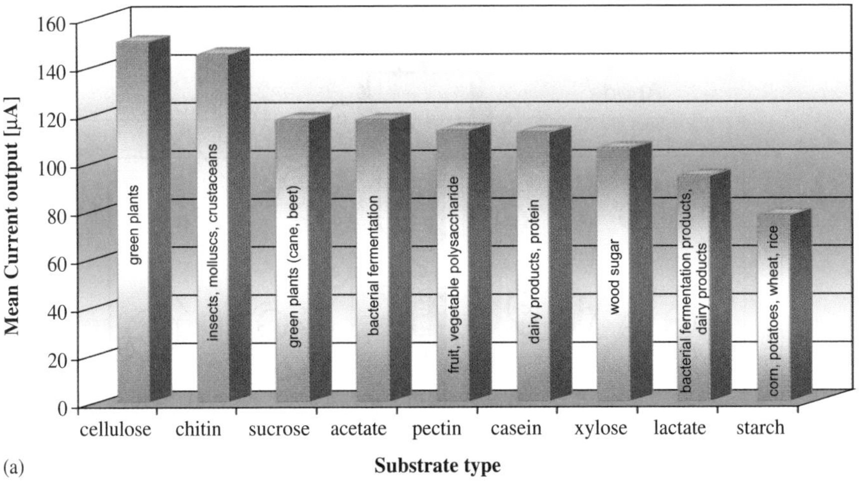
Mean Current output [μA]
160
140
120
100
80
60
40
20
0
green plants
insects, molluscs, crustaceans
green plants (cane, beet)
bacterial fermentation
fruit, vegetable polysaccharide
dairy products, protein
wood sugar
bacterial fermentation products, dairy products
corn, potatoes, wheat, rice
cellulose
chitin
sucrose
acetate
pectin
casein
xylose
lactate
starch
Substrate type
(a)

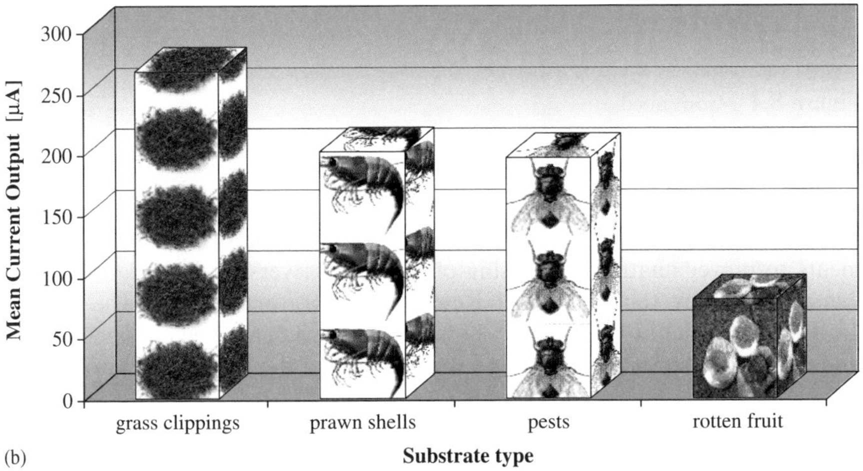
Mean Current Output [μA]
300
250
200
150
100
50
0
grass clippings
prawn shells
pests
rotten fruit
Substrate type
(b)

**Figure 8.5** a) Substrate diversity (refined organic) (courtesy of C. Melhuish, I. Ieropoulos, J. Greenman and I. Horsfield, Bristol Robotics Laboratory (BRL)). b) Substrate diversity: unrefined organic (courtesy of C. Melhuish, I. Ieropoulos, J. Greenman and I. Horsfield, Bristol Robotics Laboratory (BRL)).

chemicals [glucose, acetate and cysteine in the examples shown (purple circles)]. Not included in this figure are systems that are based on hydrogen and ferricyanide.[29]

Despite the considerable worldwide activity on the development of MFCs, the provision of sufficient energy and peak power for a SASS from a MFC represents a major challenge. To address this shortcoming, novel biological energy-generation systems need to be investigated further. One approach is to develop a system that is self-sustaining.

**Figure 8.6** The Benthic Underwater Generator (BUG) that uses marine sediment to produce peristent power for in-water operations. Top picture shows a marine buoy powered by BUG. Bottom picture shows BUG being lowered into sediment to begin energy harvesting operations. (Photographs courtesy of Lenny Tender, Naval Research Laboratory, USA.)

## 8.3 Towards Autonomy: Self-sustaining Systems

Energy self-sustaining robotic systems are being developed for use on land and in water environments. The initial steps in this direction were elegantly demonstrated in 2000 at the College of Engineering, University of South Florida. Researchers[30] developed a system called Gastonome (known as Chew-Chew) which used a fuel cell to charge Ni-Cd batteries. This novel system used *E. coli* fed with refined sugar, and a synthetic mediator (HNQ) for the electron transfer process to the chemical fuel cells (Figure 8.8).

A system called SlugBot, built by the Bristol Robotics Laboratory,[31] demonstrated that it is feasible to build a system that could search for and

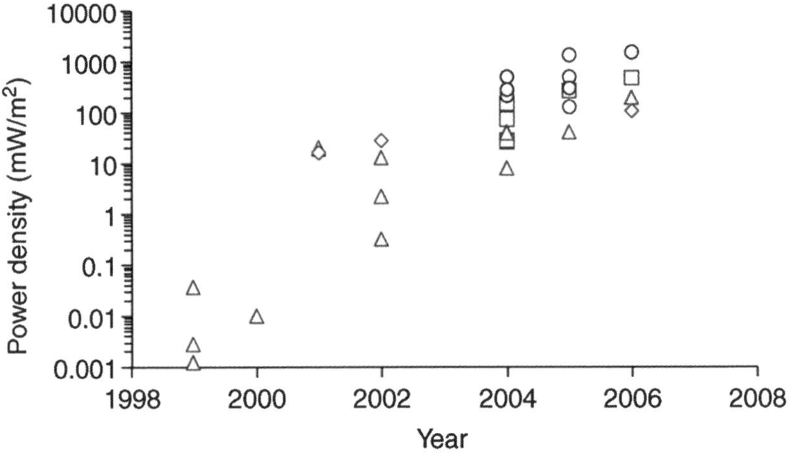
10000
1000
100
10
1
0.1
0.01
0.001
Power density (mW/m²)
1998
2000
2002
2004
2006
2008
Year

**Figure 8.7** Microbial Fuel Cells (MFC's) Power Densities[29] with permission from Elsevier. (Colour coding is given in the text).

USF

**Figure 8.8** Gastrobot Chew-Chew (Photograph courtesy of S. Wilkinson, University of South Florida, USA).

discriminate from the background a potential food source. In this case, the robotic system was programmed so that it could discriminate a plastic slug from the background of soil, use a mechanical grabbing device fitted with an infra-red seeker and collect the slug and put it into a digester. This

(a)

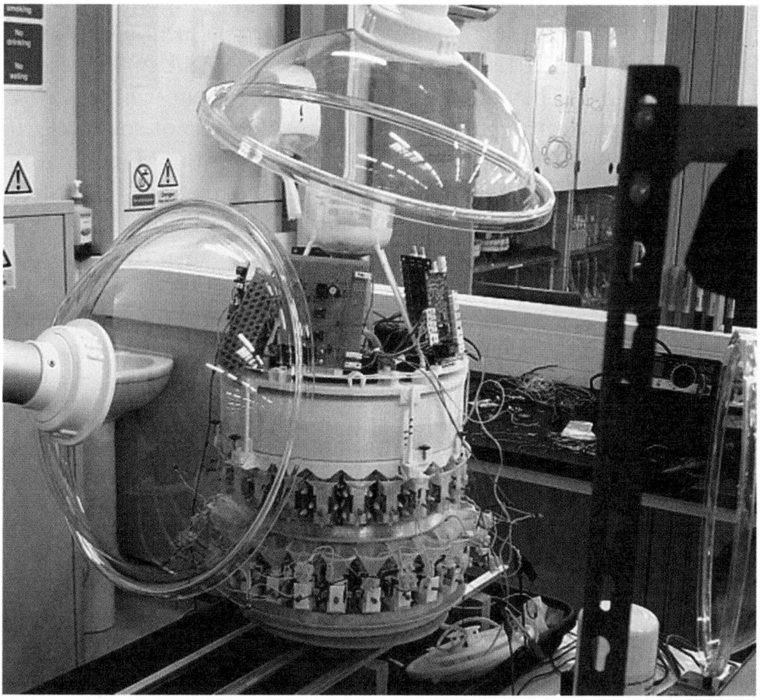

(b)

**Figure 8.9** a) Eco Bot II. World's 1st robotic system to exhibit 4 kinds of behaviour: sensing, processing, actuation (phototaxis) using $O_2$ cathode MFC's utilising unrefined substrate of flies and rotten fruit. (Photograph Courtesy of I. Ieropoulos and C. Melhuish, Bristol Robotics Laboratory (BRL)). b) Eco-Bot III. A Robotic system that can ingest fresh fuel from the environment, digest and egest waste (Photograph courtesy of I. Ieropoulos and C. Melhuish, Bristol Robotics Laboratory (BRL)).

demonstrated the potential for a robotic system to become self-sustaining within the environment. The concept has been further refined in the EcoBot series, which demonstrates that systems in the future could rely on harvesting energy from the environment. In Figure 8.9(a), EcoBot II is shown. It has eight $O_2$ cathode MFCs containing sewage sludge microbes using raw unrefined substrate *e.g.* flies or rotten fruit to generate power. It is the world's first robotic system to exhibit four kinds of behaviour: sensing, processing, actuation (phototaxis), communication, utilizing MFCs.[23,32] In Figure 8.9(b), EcoBot III is shown. It is designed to exhibit true autonomy. In its behavioural repertoire it is programmed to ingest fresh fuel from its environment, digest it, maintain a periodic fluid flow onboard and eventually egest waste.[33]

## 8.4 Space Exploration

One vivid example of how critical power and energy are to sustaining remote operations can be seen with the space programme. Autonomous robotic systems such as probes and satellites used for space applications will depend on gathering energy from the sun or on-board energy production or storage. For example, the Mars Science Laboratory,[34] which is a NASA surface exploration rover, will land on the planet Mars in 2012. The system will be powered by radioisotope thermoelectric generators (RTGs). The use of solar cells to provide power for the unit is regarded as impractical for the Mars environment. This is especially true at the higher latitudes of the planet due to shaded areas, dust clouds and the night periods when power would be drained to keep the systems warm. Hence, RTGs will be used as they can provide reliable and continuous power during daylight and night-time operations. Whilst this type of approach is practical for space application, the practicality and safety issues associated with RTGs preclude their use in a wider range of remote deployed sensors on earth. The planned RTG output is 125 watts falling to 100 watts after 14 years of operation.[35] The rover has a power management system which powers a highly functional platform with cameras, sample analysis systems and environmental sensing systems. At the moment exploratory systems such as satellites, space laboratories and exploration rovers will continue to rely on solar cell arrays and RTGs for power. However, space exploration systems will eventually utilize other forms of energy available in the universe to power systems.

## 8.5 Conclusion

Within the bio-inspired context, energy and power are commodities that can and will limit the scope (size) and availability of sensors and sensing systems. The design of any sensor or sensing system, including the conceptual SASS, must take this limitation into account. For instance, it may be that on a deployed sensing system all sensors cannot be up and running all the time due to power limitations. As envisioned in an autonomous platform like the SASS, the system, if designed properly, can reconfigure itself to maximize the usefulness of input data from the sensors that are powered up. When the analyte of interest is detected, the system can then start to power up the 'sleeping' sensors.

One of the greatest challenges with respect to sensing systems will be the supply and management of power. Each device will have its own power requirements. They can range from low voltage and current supply over an extended period of time to high voltage and current supply over a very short period of time. Efficient storage and use (and therefore power management) will play a significant role in addressing power requirements.

# References

1. I. Albinsson, S. Björkert, T. Gustafsson, B.-E. Mellander, S. J. Savage and J. Thomas, ENERGEN – Energy generation and storage for the dismounted warrior, FOI-R-2444-SE Dec. 2007.
2. B. Morgan, S. Bedair, J. S. Pulskamp, R. G. Polcawich, C. Meyer, C. Dougherty, X. Lin, D. Arnold, R. Bashirullah, R. Miller and M. Roosz, *Power considerations for MAST platforms*, 2010, *Proc. SPIE*, 7679.
3. www.biopowersystems.com and www.eurekamagazine.co.uk/article/8054/sharks-and-seaweed-inspire-green energy.aspx
4. www.qinetiq.com/home/products/zephyr.html
5. O. Kruse, J. Rupprecht, J. H. Mussgnug, G. C. Dismukes and B. Hankamer, *Photchem. Photobiol. Sci.*, 2005, **12**, 957–970.
6. P. H. Raven, R. F. Evert and S. E. Eichhorn, *Biology of Plants*, Freeman and Company Publishers, New York, 7th edn, 2005, pp. 124–127, ISBN 0-7167-1007-2.
7. B. Esper, A. Badura and M. Rogner, *Trends Plant Sci.*, 2006, **11**(11), 543–549.
8. www.technology.newscientist.com/article/dn13344
9. www.innovations-report.com/html/reports/life_sciences/report-91706.html
10. www.als.lbl.gov/als/science/sci_archive/207biomimetic-dye.html
11. www.newstarget.com/z021336.html
12. www.newscientisttech.com/channel/tech/mg19325966.600
13. W. R. Grove, *Phil. Mag. J. Sci.*, 1839, **XIV**, 127–130.
14. J. Larminie and A. Dicks, *Fuel Cell Systems Explained*, Wiley-Blackwell, Chichester, 2nd edn., 2003.
15. R. O'Hayre, W. Colella, S. Cha and F. B. Prinz, *Fuel Cell Fundamentals*, John Wiley & Sons, New York, 2009.
16. H. P. Bennetto, G. M. Delaney, J. R. Mason, C. F. Thurston, J. L. Stirling and D. R. DeKeyzer, US Pat. 4,970,145, 1990.

17. H. P. Bennetto, *Bio. Tech.*, 1990, **1**(4), 163–168.
18. K. Rabaey and W. Verstraete, *Trends Biotechnol.*, 2005, **23**(6), 291–298.
19. I. A. Ieropoulos, J. Grenman, C. Melhuish and J. Hart, *Enzym. Microb. Tech.*, 2005, **37**(2), 238–245.
20. I. A. Ieropoulos, C. Melhuish and J. Greenman, *Energetically Autonomous Robots, Intelligent Autonomous Systems*, ed. F. Groen, N. Amato, A. Bonarini and E. Yoshida, IOS Press, Amsterdam, vol. 8, 2004.
21. I. H. Ieropoulos, J. Greenman, C. Melhuish and J. Hart, *J. Power Sources*, **145**(2), 253–256.
22. I. A. Ieropoulos, C. Melhuish and J. Greenman, *Bioinspir. Biomim.*, 2007, **2**, S83–S93.
23. C. Melhuish, I. Ieropoulos, J. Greenman and I. Horsfield, *Auton. Robot.*, 2006, **21**, 187–198.
24. L. M. Tender, C. E. Reimers, H. A. Stecher, D. E. Holmes, D. R. Bond, D. A. Lowy, K. Pilobello, S. J. Fertig and D. R. Lovley, *Nat. Biotechnol.*, 2002, **20**, 821–825.
25. B. Simmons, private communication, Sandia National Laboratory, Livermore, CA, 2005 and future.iftf.org/2004/06/the_body_electr.html.
26. C. Kompis and S. Aliwell (ed.), *Sensors and Instrumentation Knowledge Transfer Network Report*, June 2008, Energy harvesting technologies to enable remote and wireless sensing, http://server.quid5.net/~koumpis/pubs/pdf/energyharvesting08.pdf.
27. J. G. De Steese, D. J. Hammerstrom and L. A. Schienbein, Pacific Northwest National Laboratory PNNL-13336, Sept 2000, Electrical Power from Ambient Energy Sources.
28. www.physorg.com/news142180005.html
29. B. E. Logan and J. M. Regan, *Trends Microbiol.* 2006, **14**(12), 512–518.
30. S. Wilkinson, Gastronome – A Pioneering Food Powered Mobile Robot, Proceedings of the IASTED International Conference on Robotics & Applications, RA318-037,176-181, Honolulu, Hawaii, August 14–16 2000 (www.gastrobots.com)
31. I. Kelly and C. Melhuish, A Slug Detection System for the Slugbot, Proceedings of Towards Intelligent Mobile Robots (TIMR), Manchester, April, 2001, Technical Report Series, Manchester University, Department of Computer Science.
32. I. Ieropoulos, C. Melhuish, J. Greenman and I. Horsfield, *Int. J. Adv. Robotics Sys.*, 2005, **2**(4), 295–300.
33. I. Ieropoulos, J. Greenman C. Melhuish and I. Horsfield, EcoBot-III; a robot with guts (in press), accepted for *The Proceedings of the 12th International Conference on the Synthesis and Simulation of Living Systems (A life XII)*, 19–23 Aug 2010, Odense, Denmark.
34. Wikipedia.org/wiki/Mars_Science Laboratory.
35. A. K. Misra, Overview of NASA Program on Development of Radioisotope Power Systems with High Specific Power, 26-06-2006, NASA/JPL.

CHAPTER 9

# *Processing and Communications*

## 9.1 Introduction

In Chapters 5 through 7, the collection, sampling and transduction of sensory stimuli were discussed. To be useful, the information arising from the sensory stimuli must be processed and ultimately acted on or sent on for further processing. Processing and communication of sensory input involves internal actions such as the fusion of data from a number of sources. These sources may be an integral part of the organism or device or may be from other organisms or devices. The result of processing the sensory input may be an action or response or may result in communication of the results to others.

A classic example of processing and response is the flight path of a moth towards a source of a pheromone. The moth senses the pheromone and adjusts its flight path towards the perceived source. If the pheromone becomes more concentrated, the moth continues on the path towards the perceived source. If the pheromone becomes less concentrated, it adjusts its flight path to find a space with a higher concentration. This sense and adjust cycle continues until the moth arrives at the source. There are also collective behaviours found in Nature that involve processing and communication of sensory input. Swarming, shoaling and the formation of flocks are examples of behaviours that require the processing and communication of sensory input from others involved in these collective behaviours (Figure 9.1).

Parameters involved in processing and communication in biological systems are shown schematically in Figure 9.2. They will be discussed in this chapter.

Bio-inspired Materials and Sensing Systems
By Peter Biggins, John Hiltz and Anne Kusterbeck

Published by the Royal Society of Chemistry, www.rsc.org

**Figure 9.1** Collective behaviour of ants leads to the formation of a bridge between leaves. This requires communication and processing of sensory data.[19]

## 9.2 Processing and Communication

Processing of sensory input and communication can be thought of in two ways. One is internal and involves data fusion, cognition, possible actions or responses that require communication between the processor and the actuator (for instance muscles). The second involves communication or transmission of data to others. This is external communication.

In human beings, the processing of sensory input is done in the brain, a processor of unsurpassed capabilities. The brain routinely processes visual, auditory, olfactory, haptic and thermal input and activates a broad range of responses. These responses have a contextual component. Fire provides visual, auditory, olfactory and thermal inputs. If a person is at home and there is a fire in a fireplace, they might move closer to the fire to enjoy its warmth. If they become too warm they will move away from the fire. Here processing of the sensory input has led to an action. A person might also indicate (communicate) to others that the room is warm and cosy or that sitting too close to the fire is uncomfortable. However, if the person is in a structure where open fires are not expected, the

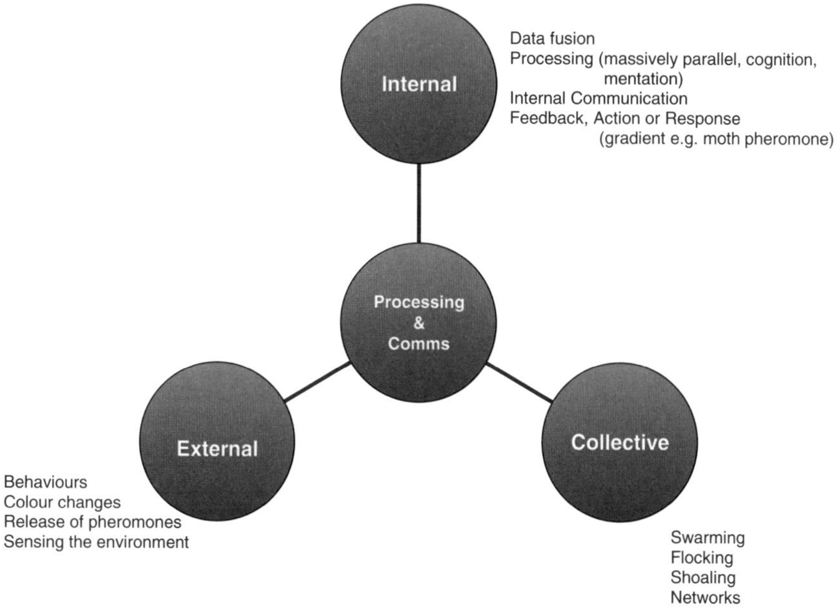
Internal
Data fusion
Processing (massively parallel, cognition, mentation)
Internal Communication
Feedback, Action or Response (gradient e.g. moth pheromone)
Processing & Comms
External
Behaviours
Colour changes
Release of pheromones
Sensing the environment
Collective
Swarming
Flocking
Shoaling
Networks

**Figure 9.2** Parameters involved in processing and communication in natural systems.

same inputs will result in a different response. These might include pulling a fire alarm, calling for help, making an effort to put out the fire or leaving the space and calling the fire services for assistance with suppressing the fire. But how do biological systems process sensory input? And, maybe more importantly, can knowledge of how this is done be used to guide the design of bio-inspired processing devices? These questions will be discussed in the next sections.

### 9.2.1 Parallel Computing

Computing has developed rapidly over the last 50 years. During most of this time computers have been built using the von Neumann architecture, that is, they have a single processor and can perform one task at a time. As more powerful processors were developed, the speed of computation increased and the time required to complete each task decreased. Although increases in processor speeds have reduced the time to complete calculations and solve problems, the one-task-at-one-time architecture of these systems is limiting.

More recently there has been a move to parallel computing. This allows more than one task to be completed at one time by using processors working in parallel. The use of multiple processors working in parallel has significantly reduced the time to complete very complicated calculations. However, the results expected from parallel computing have not been fully realized. Flynn[1] feels this is due to the difficulty in finding large and consistent degrees of parallelism within a single programme because the partitioning of tasks is too difficult. The development of new algorithms for parallel computing or tools that can decompose existing algorithms into those enabling the completion of many concurrent tasks are required for parallel computing to be successful. Another way of looking at this is that the problem to be solved must be represented in a way that allows optimum use of the parallel computing.

Natural processing systems provide inspiration for the development of massively parallel computing. These will be discussed in the following section.

## 9.2.2 Natural Computing

Natural computing is a field where models and computational techniques inspired by Nature are investigated and attempts are made to understand the world in terms of information processing. Natural computing is a highly interdisciplinary field that connects the natural sciences with computer sciences. In a review of computing paradigms inspired by natural systems, Kari and Rozenberg[2] note that inspiration for natural computing has been taken from a very diverse range of phenomena including self-reproduction, the functioning of the brain, Darwinian evolution, group behaviour, the immune system, cell membranes and morphogenesis. These paradigms can be used either on electronic hardware or on physical media such as DNA or RNA computing devices.

Cellular automata, neural computation, evolutionary computation, swarm intelligence, immunological computation, membrane computing and amorphous computing are examples of computational models inspired by Nature. These are discussed in reference 2. Amorphous computing is described in more detail in the following section.

### *9.2.2.1 Amorphous Computing*

Amorphous computing and cellular computing are examples of vastly parallel computing. Each device (computer) is simple, may be unreliable and can communicate with only a small number of its nearest neighbours. To take advantage of this system requires the development of paradigms and programming language tools that exploit the ability of vast parallelism.

Amorphous computing is an approach to computing inspired by cellular cooperation in natural biological systems. It depends on a vast number of simple processors that are connected in unknown ways to produce a coherent behaviour.[3,4] The goal of amorphous computing is to identify organizing principles and create programming technologies that enable the creation of intentional, pre-specified behaviours from the cooperative interaction of a large number of processors.

Conceptually, the amorphous computer encompasses a collection of computational devices (particles) that are located in no fixed pattern on a surface or in a volume. The devices have no knowledge of their orientation or position, may not be working properly, may contain sensors and actuators and may be capable of causing some actions to take place. They also have some computing power and a small amount of memory. The devices, which are programmed identically, are not synchronized but are capable of operating at similar speeds as they are all produced in the same way. Each device can store a local state and can generate random numbers. Some of the devices may be initialized to particular states. Communication, which might be *via* electromagnetic radiation or chemicals depending on the nature of the device, is unreliable and limited to a few nearby neighbours. When communicating with other devices, the sending device has no assurance that the message has arrived at a receiving device. Communication can take place over a distance that is small compared to the size of the computer.

The amorphous computer has a very large number of devices (from millions to trillions) and is therefore a massively parallel computer. The algorithms required to run the computer should be independent of the number of devices and the performance of an amorphous computer should degrade as the number of devices is decreased.

The importance of this approach arises from the emergence of new technologies, such as microfabrication and cellular engineering, which are enablers for the development of amorphous computing. The simple processors (devices) arising from these technologies could act as the basis for information processing systems with immense computational power and low cost. For instance, microfabrication techniques have allowed the low-cost production of micro electromechanical devices that integrate logic circuits, sensors, actuators and communication capabilities on a single chip (device) while cellular engineering has provided the possibility of making digital logic circuits within individual cells. As cells can reproduce themselves, processing units (devices) based on cells have the potential to be produced in large numbers at very low cost.

### *9.2.2.2 Programming Amorphous Computers*

The challenge is to programme such systems and this will require the development of new programming methods. The methods must work without precise control over the interaction or arrangement of the individual computing devices, that is, within the amorphous computing model. This is in contrast to digital computers where the precise arrangement of reliable parts has been the rule.

Bio-inspired metaphors have been a source for amorphous computing programming techniques. The genetic programmes encoded in cells result in the formation of well-defined shapes and forms even though the arrangement and number of individual cells are variable. Diffusion is another natural metaphor that has been used to develop programmes for amorphous computing. In this approach, one device is chosen to broadcast a message to its neighbours, who in turn broadcast to their neighbours and so on. This results in a wave moving through the computer. The message contains a count and each device stores the received count and increments it by one prior to resending the message. Each device stops sending the message after it has stored its count and ignores subsequent count messages. This wave gives each device a rough idea about how far it is from the message source. It is also possible to define regions of controlled size by programming the devices to stop resending messages once a stored count number is exceeded.

### *9.2.2.3 Paradigms for Amorphous Computing*

Although amorphous computing is in its infancy, there are a number of programming paradigms that have demonstrated the viability of this approach even when the individual devices are unreliable and lack precise organization. These include Microbial Colony Language,[5] Growing Point Language,[6] origami-based self-assembly,[7] dynamic recruitment,[8] growth and regeneration,[9] and abstraction to continuous space and time.[10,11]

In the dynamic recruitment paradigm, processes recruit computational devices (particles) from an amorphous computer to implement their goals. Dynamic recruitment has been used to create a storage system for streaming audio and images. The fragments of the audio and images move freely within the amorphous computer and are called to a port when needed. Robustness is enhanced by storing multiple copies of the low-resolution audio or video and fewer copies of higher resolution audio or images. Although lost fragments will degrade the stored audio or images, the audio or images will not be lost. The details of the other paradigms can be found in the references above.

To take full advantage of the amorphous computing paradigms, the development of a linguistic framework is also required. The basic components of this new language are referred to as primitive mechanisms. A primitive mechanism specifies the behaviour of the individual devices in the amorphous computer. These mechanisms have names such as gossip, random choice, fields and gradients. Gossip is a simple communication mechanism that is used by a number of devices to come to an agreement on the value of a parameter, for instance, a minimum or a maximum value. Random choice is used to break symmetry and allows devices to differentiate themselves from other devices in that region of the amorphous computer. A field is associated with every component of the state of the devices in an amorphous computer over the space occupied by those devices. Gradients can estimate the distance from each device to the device indicated as the source.

The programmes for amorphous computing require rules for the combination of primitive mechanisms. Combinations of primitive mechanisms produce more complex behaviours. Means of abstraction are also required so that the compound behaviours arising from combinations of primitives can be manipulated as units.

The natural computing techniques referred to in this section have been primarily implemented on electronic hardware. However, there are other techniques that are or might be implemented on completely different types of hardware. These techniques include molecular computing and quantum computing.

## 9.3 Molecular Computing

Molecular computing is based on the concept of using biomolecules, such as DNA (deoxyribonucleic acid) strands, to encode data and biochemical tools to transform the stored data to perform arithmetic or logic functions. Adleman[12] was the first to demonstrate the use of molecular computing. He solved a Hamiltonian Path Problem using DNA strands in a test tube.

Molecular computing uses operations such as cutting and splicing of DNA strands, the synthesis of DNA strands, the reading of the nucleotides that make up a DNA strand and the synthesis of multiple copies of DNA strands to perform computations. Examples of the use of this approach to build an autonomous molecular computer include the implementation of computational state transitions using biomolecules[13] and a DNA implementation of an automaton with possible applications to the design of novel drugs.[14] However, the major contribution of molecular computing has been to provide insights into the understanding of molecular self-assembly.

## 9.4 Cognition

The development of artificial intelligence for machines and sensing systems is a goal that has not been fully realized. Ideally, one would like to develop systems that can utilize sensory input to build knowledge bases and then use this knowledge to act on the input. Actions might include movement towards or away from the source, collecting more input or communicating the data to other systems. The obvious inspiration for such a system is the human brain. However, to use the human brain as inspiration for the development of artificial intelligence requires an understanding of how processes such as cognition take place in the brain. Human intelligence involves distinct but highly interacting brain structures. For cognition these structures are the cerebral cortex and the thalamus.

A theory has been developed to explain cognition in the human brain.[15,16] It is hypothesized that cognition is carried out by thalamocortical modules. There are approximately 2000 modules and each consists of small and localized patches of the cortex that are uniquely paired with areas of the thalamus. These patches of the cortex and thalamus are connected *via* axons. Each of the modules is hypothesized to be controlled by a single excitatory control input and is responsible for giving effect to a list of symbols. The module has many thousands of symbols that represent an object's attributes. The symbols in a particular module are represented by a collection of neurons, and a collection of neurons is specific to each symbol. However, some neurons may be present in more than one symbol and a particular neuron may participate in many symbols. These symbols are terms of reference for describing objects in the mind and for carrying out a single symbolic processing operation called confabulation.

Confabulation is put into action by neurons in a particular module. The process takes tens of milliseconds and involves a winner-takes-all type of competitive interaction between symbols. The states of the neurons in a module evolve during confabulation and involve a large number of parallel interactions between the neurons in that module. The final state of each involved neuron is either excited (active) or inactive. If the output of confabulation is a single symbol, all the neurons involved with that symbol become active and all other neurons in that module become inactive. If more than one symbol is the result of the confabulation, the neurons in these symbols will have different levels of excitation. The winning symbol (or symbols) is referred to as a conclusion. It is possible that no neurons are excited in a confabulation and this indicates that no conclusions were reached. This is equivalent to humans being asked a question and responding that they do not know the answer.

The theory further hypothesizes that for knowledge to be stored in the thalamocortex and used for cognition requires axonal connections between neurons for a symbol in one module and neurons for another symbol in another module. These connections or links are referred to as items of knowledge. Humans are thought to have billions of items of knowledge resulting from these links. If a confabulation results in the expression of a single symbol, then an action command associated with that symbol is issued. Action commands may cause a muscle or thought control signal to be sent that may eventually result in the execution of an action.

A hierarchical structure is found in most cognitive functions. That is, cortical modules are arranged into hierarchies and a symbol in a higher level module represents many combinations of symbols at a lower level. The structure of the modules is such that symbols for commands that would conflict with one another are grouped in the same module. As only one action signal can be issued by the module at any one time, this eliminates conflicting actions.

The theory is defined in more detail in reference 16 and the author notes that the theory is an initial plan for mechanizing cognition. A limitation of confabulation theory is that it is restricted to knowledge development using formatted processes. It cannot mimic the dynamic, autonomous and satisfaction driven process used by the brain.

## 9.5 Applications

### 9.5.1 Sensor Networks

The amorphous computing paradigm is applicable to sensor networks. The individual sensors are the equivalent of the individual computing devices in an amorphous computer. The sensors could be randomly distributed in a space or volume and have rather simple capabilities. However, they would require the ability to communicate with other sensors in their vicinity and pass on information about their environment *via* a number of routes to other sensors. Processing of information from an array of sensors would be used to develop an accurate picture of the environment being sensed. The large number of sensors and the ability to self-heal (establish new links) when devices fail would make such a system extremely robust. The rather simple capabilities of the individual sensors would reduce the cost of the sensor network.

## 9.6 Insect Sensory Systems

Insect sensory systems that can act as inspiration for processing and communication components and systems have been reviewed.[17] Although insects have minimal learning capability, they have been very successful because of the high level of organization in their sensory and neuromotor systems. These systems act as inspiration for the development of sensing systems for several reasons. The first is simplicity. Many insect sensory receptors are single cells and the resulting simplicity makes them easier to study than those in vertebrates. The highly distributed nature and fault tolerance of their sensory systems make them excellent models for the development of novel communications and computing systems. In addition, insects assembled into populations demonstrate the characteristics of more complex systems including self-organization, emergent behaviour and distributed function. These attributes are applicable to the design of sensor networks where the ability to build reliable and adaptive networks from simple components is desirable.

Insect sensory systems perform a number of significant computing and information-processing functions. The chemical sensory systems of insects are a case in point. Insects use chemicals referred to as semiochemicals or infochemicals to both encode and decode messages. The chemicals include both pheromones and allelochemicals and enable communication that is specific, sensitive, reliable and robust. Chemical-based insect sensory systems provide inspiration for novel man-made wireless networks. There has also been considerable research carried out on other insect sensory systems including hearing and mechanosensation, vision, gustation (taste) and olfaction.

### 9.6.1 Collision Avoidance/Motion Detection Systems

Insect vision systems have been used as inspiration for motion detection systems. These include sensors for robots and analogue very large scale integration (VLSI) chip designs. The lobula giant movement detector (LGMD) and descending contralateral movement detector (DCMD) neurons of locusts are capable of detecting objects, such as predatory birds, that are on a collision course with them. The neurons respond to objects on a collision course by producing responses that increase in frequency as the objects moves closer. The LGMD then processes the excitation from the neurons related to the increase in size of the object as it moves closer. The processing removes much of the excitation caused by stimuli such as the background and rotational movements that are not coming from objects on a collision course with the locust.

Models of the LGMD have been used to detect collisions in robots and in real collision scenes. However, these models could not differentiate between objects on a collision course and objects moving across the field of view of the neurons in the detector. To address this problem, the LGMD model was modified. This modification was based on directionally sensitive neurons called elementary movement detector (EMD) neurons that are found in some insects.[18] The combination of EMD neurons with the prefiltering of input impulses by the LGMD

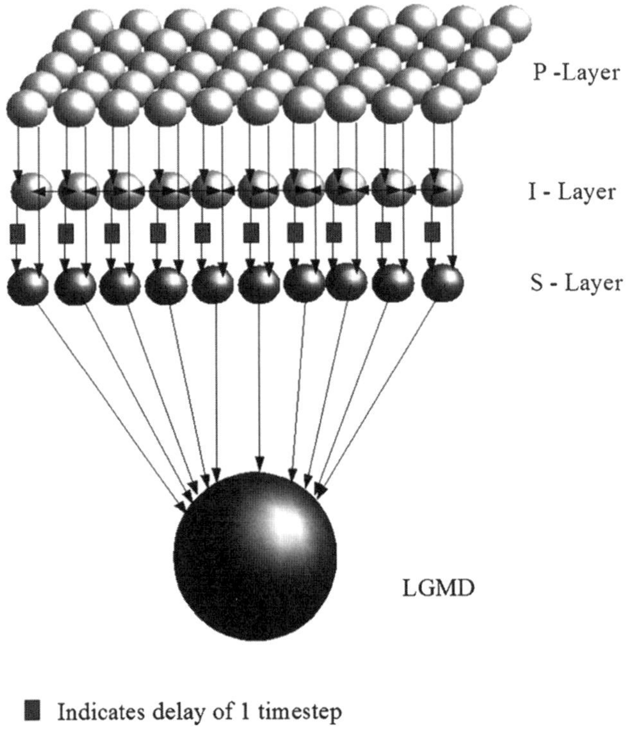
P -Layer
I - Layer
S - Layer
LGMD
Indicates delay of 1 timestep

**Figure 9.3** The lobula giant movement detector (LGMD) and its input afferents. The greyscale value of each pixel is mapped onto the corresponding photoreceptor unit (P-layer). The output of the photoreceptors is the change in greyscale value of the pixel between successive time steps. This output is passed to the retinotopic excitatory S-layer and the inhibitory I-layer. The inhibition in the I-layer spreads between neighbouring I units to spatially blur the change in image passed out of the photoreceptor units. The inhibition is delayed by one time step and is subtracted from the excitation at the S units. Resultant excitation is passed to the LGMD where spikes occur if the excitation exceeds a given threshold.[18] (Reprinted from R. Stafford, R. D. Santer and C. Rind, *Biosystems*, 2007, **87**, 164–171, copyright 2007, with permission from Elsevier.)

resulted in a collision detector system that can distinguish between objects that are translating from one side of the detector's field of view to the other and those on a collision course with the detector. The LGMD and EMD portions of the collision detection model are shown schematically in Figures 9.3 and 9.4.

Inputs from EMDs, which take information from the centre of the image that is being processed, are fed into four directional sensitive movement detectors (DSMDs). These DSMDs can detect either vertical movement (up and down) or horizontal movement (left and right). The vertical and horizontal DSMDs take input from a column and a row of ten EMDs respectively located at the centre of the $150 \times 100$ pixelated image.

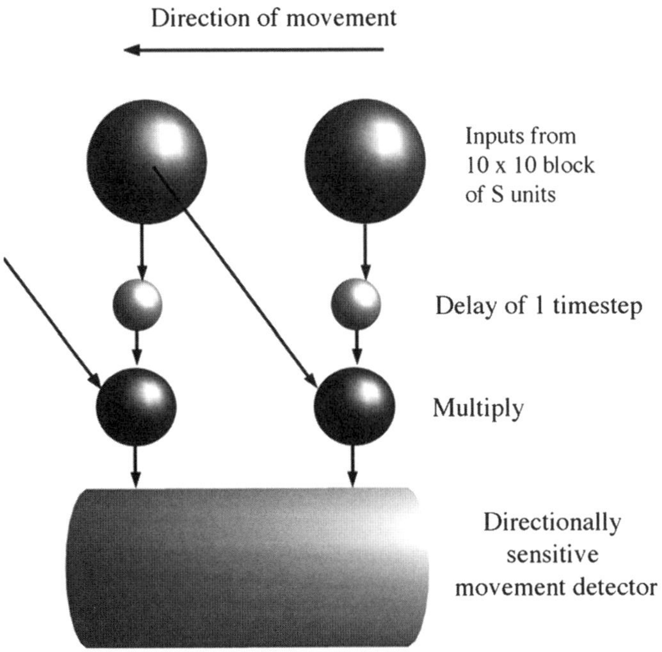
Direction of movement
Inputs from
10 x 10 block
of S units
Delay of 1 timestep
Multiply
Directionally
sensitive
movement detector

**Figure 9.4** The elementary movement detector (EMD) unit. Excitation is obtained by each half of an EMD from summing a group of $10 \times 10$ S units in the LGMD. In this left movement detecting EMD excitation in the right-hand half of the EMD is passed retinotopically with a delay of one time step to a multiplication unit. The multiplication unit also receives excitation with no delay from the left half of the EMD. If movement leftwards has occurred then excitation should be present from both halves of the EMD at the multiplication unit, if not then at least one of the excitations should be close to zero. The result of the multiplication is passed to a directionally sensitive movement detecting neuron, Hl.[18] (Reprinted from R. Stafford, R. D. Santer and C. Rind, *Biosystems*, 2007, **87**, 164–171, copyright 2007, with permission from Elsevier.)

## 9.7 Bio-inspired Networking

As networks expand, problems and challenges associated with increased network complexity, resource constraints, heterogeneous architectures, lack of centralized control and infrastructure, the dynamic nature of these networks and the requirement for robustness (survivability) must be addressed. Natural networking and communication systems have evolved that can deal with changing environmental conditions, and are self-healing (resilient to damage and failures), resource efficient, capable of collaborative operation using a very simple set of rules, self-organizing and have capabilities that are greater than the sum of those of the individual components of the system. As these challenges have been dealt with by natural systems, there is much that can be learnt from Nature concerning approaches to networks and communication. An excellent review of how natural systems can act as an inspiration for approaches to the development of novel networking technologies has been published.[19]

To be used to inspire approaches to networking and communication, an understanding of how natural systems function is required. This understanding can then be used to design paradigms that can be used to develop novel systems that are able to adapt to changing environments, and are heterogeneous, self-organizing and scalable.

Ant colonies exhibit behaviours that are the basis of solutions to several of the challenges associated with large networks. The foraging behaviour of an ant colony is an example of the use of limited resources to achieve an optimized behaviour for the ant colony. Ants forage for food using a random-walk approach. When an ant finds food it returns to the colony by following its own path and releases pheromones to mark the path. Other ants detect the pheromone and follow it to the food source. As the food source is depleted, ants stop returning to the colony, the pheromones evaporate and ants continue with the searching process. Even when food is found some ants will continue with the search process. This optimizes the search strategy by ensuring that new sources of food continue to be found. This behaviour is the inspiration for resource efficient networking technologies.

### 9.7.1 Bio-inspired Network Routing Protocols

The behaviour used by ant colonies to optimize their search for food is one of the best analysed examples of swarm intelligence. An ant colony optimization (ACO) algorithm has been described.[20,21] The most important aspect of this algorithm is the transition probability for an ant to move from one position to another. This represents the routing information for the foraging process and depends on a list of positions or sites not visited by the ant, the distance between the ant and a site, the concentration of a pheromone on the path between two sites and adjustable parameters that control the weight of the distance and pheromone level parameters. The quantity of pheromone that an ant lays following a tour on a line between two sites can also be described mathematically. Changes in the environment are also important, for instance, the rate of evaporation of the pheromone. If ants become unsuccessful, the pheromone disappears and the ants continue looking for food. The number of ants is also critical to the approach providing optimal solutions. If there are too many ants then suboptimal paths will be selected and this leads to less than optimal solutions.

Ant colony behaviour (optimization) has been used to design routing protocols for networks. One of these, AntNet,[22] periodically sends mobile agents to randomly selected destination nodes to determine a minimum cost path between the source and the destination nodes. Once these mobile agents locate the destination node they travel back on the same path to the source. As this happens, the routing tables of the traversed nodes are updated with the most recent information on the destination node. To avoid congestion on the

**Table 9.1** Current research projects on bio-inspired networking.

| *Project name* | *Funding* | *Research area* | *URL* |
|---|---|---|---|
| ANA | EU FET | Autonomic network architecture and principles | http://www.anaproject. org/ |
| BioNet | NSF, DARPA | Bio-networking architecture for design and implementation of scalable, adaptive, survivable/available network applications | http:// netresearch.ics.uci.edu/bionet/ http://www.bionets.eu/ |
| BIONETS | EU FET | Bio-inspired service evolution for the pervasive age, | http://www.cascadasproject.org/ |
| CASCADAS | EU FET | autonomic and situation-aware communications and dynamically adaptable services | |
| *ECAgents* | EU FET | Embodied and communicating agents interacting directly with the physical world | http://ecagents.istc.cnr.it/ |
| Haggle | EU FET | Situated and autonomic communications | http://www.haggleproject.org/ |
| *MC* | NSF, DARPA | Molecular communication as a solution for communication between nanomachines | http://netresearch.ics.uci.edu/mc/ |
| *Swarmanoid* | EU FET | Design, implementation and control of a novel distributed robotic system | http:// www.swarmanoid.org/ |
| *Swarm-bots* | EU FET | Design and implementation of self-organizing and self-assembling artefacts | http://www.swarmbots. org/ |
| *WASP* | EU IP | Self-organization of nodes and services in Wireless Sensor Networks (WSNs) | http://www.waspproject.org/ |

network, the AntNet maintains a probability of creating new mobile agents based on the traffic conditions present on the network. As one path becomes congested, others can be established between the source and destination nodes.

There are currently a number of research projects in the area of networking that are bio-inspired. Some of these are listed in Table 9.1.

## 9.8 Issues

The processing and communication functions of a SASS have many of the issues described for the other components of a SASS. One of the major issues is interfacing the processing unit with the sensors, and communication and actuation devices that might be part of the system. For instance, the sensors might be hard wired to the processing unit or might send their input wirelessly to a processor remote from their location. The system might have the ability to move and this will require sending signals to the devices that produce locomotion. These factors have an impact on the design, size and power requirements of the device. If the processing unit can be remote from the sensors then a single processing unit can receive and process data from a number of sensing elements. This may not be preferred as the loss of the processing unit will disable a large number of sensors. However, the provision of autonomy (processing, communication and movement, *etc.*) to each of the systems may be cost prohibitive.

## 9.9 Conclusion

Although developments in amorphous computing and machine intelligence show promise, they are still in the theoretical or developmental stages. The use of a vast number of simple distributed sensors and the processing of their sensory input will be greatly enhanced if paradigms of amorphous computing, for instance, can be realized in operational systems. This will also require the development of computing languages that take full advantage of the capabilities of the distributed sensors. One can envision a robust network of inexpensive sensors with the ability to communicate with and through other sensors. These sensors could be deployed to detect chemical or biological agents in an operational area. Communication with other sensors with respect to their position and the concentration of agent would allow the mapping of the agent plume and concentration gradients and, in conjunction with meteorological information, facilitate predictions of the plume's course. If the sensors are mobile, they could be programmed to determine the concentration gradients in, and the extent of, an agent plume in the same way that a moth finds the source of a pheromone.

Although there has been a considerable amount of research and development in the field of probabilistic neural networks (PNNs), the development of machines with enhanced cognitive abilities will result in sensing systems with increased autonomy and the ability to deal with anomalous information. From the user's perspective, this will reduce false positives and may result in lower detection thresholds for the sensory input.

## References


1. M. J. Flynn, *Computer* 1996, **29**(7), 151–152.
2. L. Kari and G. Rozenberg, *Communications of the ACM*, 2008, **51**, 72.
3. H. Abelson, D. Allen, D. Coore, C. Hanson, E. Rauch, G. J. Sussman and R. Weiss, *Communications of the ACM*, 2000, **43**, 74.
4. H. Abelson, J. Beal and G. J. Sussman, *Amorphous Computing*, MIT Computer Science and Artificial Intelligence Laboratory, Technical Report MIT-CSAIL-TR-2007-030, June 1, 2007.
5. R. Weiss, PhD thesis, MIT, 2001.
6. D. Coore, PhD thesis, MIT, 1999.
7. R. Nagpal, PhD thesis, MIT, 2001.
8. W. Butera, PhD thesis, MIT, 2002.
9. A. Kondacs, in *International Joint Conference on Artificial Intelligence (IJCAI)*, Acapulco, Mexico, 2003, pp. 633–638.
10. J. Beal and J. Bachrach, *IEEE Intell. Syst.*, 2006, 10–19.
11. R. Newton, G. Morrisett and M. Welsh, in *International Conference on Information Processing in Sensor Networks*, Cambridge, MA, April 2007, pp. 489–498.
12. L. M. Adleman, *Science*, 1994, **266**, 1021.
13. K. Sakamoto, H. Gouzu, K. Komiya, D. Kiga, S. Yokoyama, T. Yokomori and M. Hagiya, *Science*, 2000, **288**, 1223.
14. E. Shapiro and Y. Benenson, *Sci. Am.*, 2006, **294**, 44.
15. R. Hecht-Nielsen, *Neural Network.*, 2005, **18**, 111.
16. R. Hecht-Nielsen, in *Biomimetics – Biologically Inspired Technologies*, ed. Y. Bar-Cohen, CRC Taylor & Francis Group, Boca Raton Florida, 2006, p. 57.
17. Z. Ma and A. W. Krings, *Ad Hoc Networks*, 2009, **7**, 742.
18. R. Stafford, R. D. Santer and F. C. Rind, *Biosystems*, 2007, **87**, 164.
19. F. Dressler and O. B. Akan, *Comput. Network.*, 2010, **54**, 881.
20. M. Dorigo, G. Di Caro and L. M. Gambardella, *Artif. Life*, 1999, **5**, 137.
21. M. Dorigo, V. Maniezzo and A. Colorni, *IEEE Trans. Syst. Man. Cybern.*, 1996, **26**, 1.
22. G. Di Caro and M. Dorigo, *J. Artif. Intell. Res.*, 1998, **9**, 317.

CHAPTER 10

# The SASS Approach

## 10.1 Introduction

Over the course of evolution Nature has developed a wide range of materials, forms and functions, and the ability to self-sustain, replicate, adapt and survive. These capabilities can be exploited for a wide range of materials and sensing applications. Whilst we can learn new paradigms from biological systems, perhaps the most remarkable aspect of Nature is its diversity and ability of species to exist within a number of very different environments. The diversity is accomplished using a limited range of materials in unique ways to design and build in function. For example, biological structures have hierarchical design and their structure is controlled on scales that range from the molecular to the macroscopic. The information required to assemble an organism's structure is encoded by its DNA. Therefore an understanding of the sequencing and timing of the assembly processes will be essential to fully appreciate how organisms in Nature carry out synthesis and assembly on a concurrent basis from within.

Proteinaceous material is the major constituent of functional biological tissues, whereas rigid composite structures found in Nature incorporate inorganic compounds such as calcium carbonates, calcium phosphates, and silica, as shown in Figure 10.1. These materials are not only multifunctional but are produced at less than 40 °C and atmospheric pressure from readily available materials in the environment. Depending on the level of complexity of the organism, the time required for it to achieve optimum functionality will vary considerably, for example, from minutes to years. The fabrication rates found in Nature may be too slow for many commercial processes but this will ultimately depend on the complexity of the final product.

Bio-inspired Materials and Sensing Systems
By Peter Biggins, John Hiltz and Anne Kusterbeck

Published by the Royal Society of Chemistry, www.rsc.org

**Figure 10.1** A Scanning Electron Microscope picture showing the intricate structure of a diatom (courtesy of Simon Smith, Dstl).

There are other examples of success in this area, as in the use of bacterial and viral expression systems, tissue culturing and biological scaffold building used for the fabrication of new synthetic materials for a wide range of applications. The ability to couple and interface organic materials with inorganic substrates results in new composite systems. Exploitation of our understanding of structure at the molecular level may result in tools that can fabricate hierarchical 3-D structures. This bio-inorganic interface is currently the focus of substantial research.

It is through the study of the mechanisms underpinning biological functions (and with respect to this book, biological sensing functions), their structures and processing systems that new components, processing and communication architectures and systems can be developed. In Chapter 3 the conceptual model for a situational awareness sensing system (SASS) was proposed. The bio-inspired approaches that can be used to develop alternative and innovative design of materials and components are covered in Chapters 4 through to Chapter 9. Thought construct models such as 'SASS' allow the exploration of how research in different fields of science and technology can be integrated at the component level and then packaged at the system level using Nature as a guide.

However, to fully exploit the potential of bio-inspired fabrication will require the continued development of enabling technologies. The rate at which this takes place will be dependent on a number of drivers including commercial incentives and the development of appropriate exploitation routes. The global investments being made in biotechnology, bio-inspiration and biomimicry (touched on in Chapter 2) reflect how the bio-inspired approach to materials, sensors and systems is growing. It is essential that manufacturers, designers, scientists, engineers and requirements personnel are involved at the conceptual stage of research and development processes aimed at the fabrication of novel materials, sensors and sensing systems. Success will require that inter-disciplinary, cross cutting teams be established that can exploit scientific advances by melding the science and technology base with design and manufacturing capabilities.

## 10.2 Design and Manufacture

One of the keys to taking advantage of bio-inspiration is the abilty to fabricate novel components or systems. Conventional manufacture uses techniques such as moulding, subtraction, pressing, and weaving to fabricate objects. These are in essence linear processes. Although a motor car is a three dimensional object, it is made up of components that have been formed through pressing, moulding or machining. This approach to manufacturing often uses materials in an inefficient way and involves the expenditure of a considerable amount of energy.

Another powerful lesson that can be learned from biological systems is the way in which they use readily available materials to fabricate structural and functional materials using a bottom-up approach at low temperatures. This is completely different to many manufacturing techniques. Therefore, to take advantage of the true potential of bio-inspired approaches for the development of materials, components and sensors, the fabrication processes will have to incorporate bioengineering or bio-inspired approaches to their design and manufacture.

### 10.2.1 Bioengineering

The revolution in gene synthesis technologies, which are enablers for the *de novo* creation of organisms from nothing but genetic code, opens up an almost limitless range of possibilities. Synthetic genomics holds great promise for the future. For example, the initial problems encountered in re-booting synthesized bacterium by researchers at the J. Craig Venter Institute (JCVI) and reported in September 2009 have been overcome.[1] On the 20th May 2010[2] they announced that they had successfully produced the first self-replicating synthetic bacterial cell. They synthesized the 1.08 million base pair chromosome of a modified *Mycoplasma mycoides*. The synthetic genome was assembled in yeast cells and grown as a yeast artificial chromosome. This synthetic genome was isolated from the yeast cell and transplanted into *Mycoplasma capricolum* recipient cells which had the genes for its restriction enzymes removed. After several days viable *M. mycoides* cells were produced; the synthetic cell is called *Mycoplasma mycoides* JCVI-syn1.0.

These researchers have therefore demonstrated that genomes can be computer designed and synthesized in a laboratory and then transplanted into a recipient cell to produce a self replicating cell. With these barriers successfully overcome, the possibilities of bioengineering are greatly expanded. New approaches to additive manufacture may allow a new range of systems to be built. Bio-robotic systems and systems that produce their constituent parts and enable self assembly are starting to emerge.

The combination of bio-informatics, computer modeling and system biology is leading the way in the field of automated molecular biological research. This may make it possible to produce SASS entities on a mass production basis. The use of robotics to carry out microbiological tasks within the laboratory is gaining considerable ground and a research group is now examining the concept of the 'robot scientist'.[3] The robotic system makes use of an iterative approach to experimentation, where knowledge acquired from a previous iteration is used to guide the next experimentation step. This is a process known as Active Learning. The system can plan its own agenda, *i.e.*, decide how best to improve its knowledge base and how to go about acquiring information. The system uses the laboratory robot to execute the experiment(s) selected as most informative; has a plate reader to analyse the experiments, generating data corresponding to the scientific observations; uses abductive logic programming to generate valid hypotheses that explain the observations; and uses these hypotheses to determine the next most informative experiment. At the beginning of any investigation, the Robot Scientist has not yet discovered any information; therefore all hypotheses are equally valid. As the directed discovery process continues, each new observation (or experiment/interpretation cycle) will invalidate some of the hypotheses, thereby excluding incorrect discoveries. The experiment selection process aims to choose the experiment most likely to refute the most hypotheses. This iterative process allows irrelevant experiments to be avoided, potentially saving both laboratory time and the cost of using unnecessary reagents and biological materials.

In the future this robotic approach for high throughput experimentation and learning may be linked to manufacturing systems such as additive manufacturing. With advances in automated research systems and the significant developments in design and manufacture, it is likely that we are starting to see the digital revolution merging with DNA synthesis approaches. As these enabling technologies mature, the bio-fabrication of the materials, components, and systems necessary for the realization of the SASS concept will become a reality.

## 10.2.2 Additive Manufacturing

Additive manufacturing (AM) is the process of making a part or product by adding successive layers of material in a relatively efficient way, such that there is little waste. Examples are polymeric, metallic and ceramic additive manufacturing, fused deposition modeling, laser sintering, inkjet printing and aerosol jet printing. A key approach to additive manufacture uses 3-D printing techniques where a three dimensional object is created by the deposition of successive layers of material. The 3-D printers are generally faster, cheaper and easier to use than other additive manufacturing technologies presently available and also offer product developers the ability to fabricate parts and assemblies from several materials with different mechanical and physical properties in a single build process. Advanced 3-D printing technologies yield models that closely emulate the look and feel and have the functionality of product prototypes. As 3-D printers become cheaper they will be accessible to a wide range of product developers, and as such allow garage level industry to develop. This field continues to develop and it is now possible to simultaneously deposit different types of materials.

Finally, whilst rapid prototyping dominates current uses, 3-D printers offer tremendous potential for mass production of materials, components and systems. This technology is currently used in the jewellery, footwear, industrial design, architecture, automotive, aerospace, dental and medical industries, however, it is currently being studied by biotechnology firms and academia for use in tissue engineering applications. This opens the possibility for this technology to be used in the fabrication of tissues and organs. Layers of living cells are deposited onto a gel medium and slowly built up to form three dimensional structures. This organ printing or computer-aided tissue engineering is an enabling technology for the mass production of SASS entities.

## 10.3 The SASS Approach

As our understanding of the advantages that can be derived from the application of bio-inspiration to materials, sensors, and sensor systems develops, technological developments arising from the exploitation of this understanding will be realized. An improved understanding can also be used to address requirements and perhaps eventually change the type of technologies currently used in the defence, security and civilian sectors. The SASS conceptual model and how it relates to a component, system and system of systems is shown in Figure 10.2.

Looking at bio-inspired approaches to materials, components and sensors for defence and security applications with detection and identification in mind, it is evident that these are applicable to a wide range of unattended sensing systems for use in the land, sea, and air environments as well as space exploration. In Figure 10.2, the schematic shows the levels within the SASS concept, in terms of the underpinning science and technology areas that contribute to the development of a system and the kinds of systems which can be networked together. In Table 10.1, a list of potential applications of elements of the SASS from the component to the system of systems level is given.

The list in Table 10.1 is only a small subset of the potential applications of the SASS approach. The SASS concept illustrates that advances in a capability

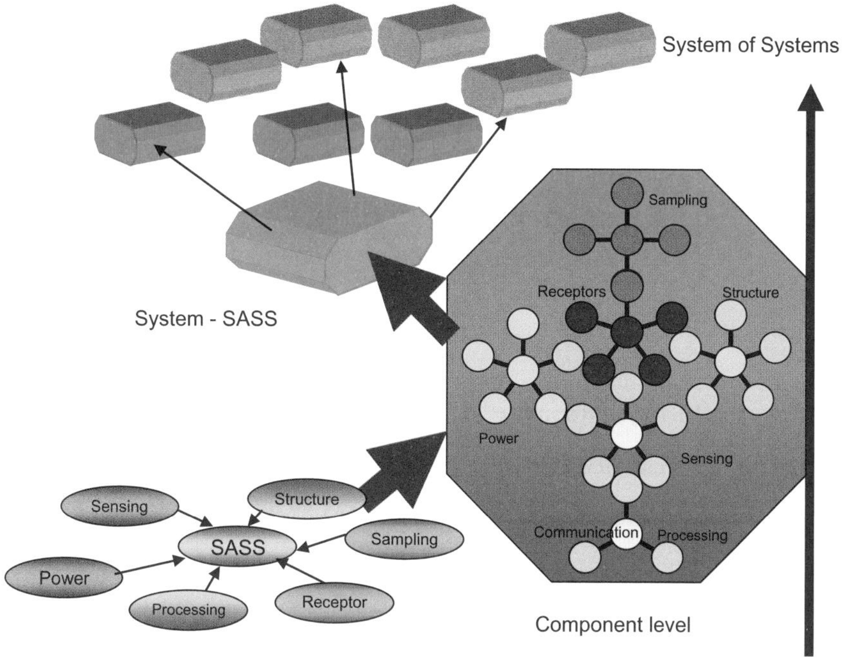
System of Systems
System - SASS
Sampling
Receptors
Structure
Power
Sensing
Communication
Processing
Component level
Sensing
Structure
SASS
Sampling
Power
Processing
Receptor

**Figure 10.2** A schematic of the levels within the SASS concept.

at the sub component and component level can be used to enhance an existing system. Similarly, at the system level the components can be used to build improved systems, for instance, a new component might enhance detection/identification capabilities or improve system mobility. In fact, there is increasing development of autonomous robotic systems for the land, sea and air environments using bio-inspired principles. An example is the use of bio-inspired vision systems based on the insects compound eye construct for navigation and guidance. Such systems can be used in urban areas for search and rescue operations, identification of wounded combatants and the surveillance of facilities.

## 10.3.1 Component Level

The transfer of knowledge from biology to technology through bio-inspiration is relevant to a wide range of applications including protection by armoured structures, camouflage, light adaptive aerodynamic structures for efficient flight, and ultra sensitive sensors. At the component level, structures and surfaces can be designed to be incorporated into detection systems that have the ability to detect toxins, chemicals, biological agents, explosives or other hazards, or for coating surfaces which have hydrophobic or hydrophilic

**Table 10.1** Potential application areas.

| *Level* | *Application area* | *Comment* |
|---|---|---|
| Component | Camouflage | Biomaterials that change colour to reflect environment |
| | Combat identification | Biological markers, pattern recognition |
| | Lightweight armour | Protective systems that self-heal, structural arrangements |
| | Biodegradable | Environment friendly material |
| | Novel processing | Protein memory/computing devices and neural architectures |
| | Biomaterials | Use of chitin and resilin for building structures |
| | Optic flow | Insect vision systems for non-GPS and visual resolutions |
| | Radiation resistance | Protein-based components for electronic circuits |
| | Vaccine development | New ways of developing countermeasures for diseases |
| | Wound healing | Tissue/organ growth, accelerated healing regimes |
| | Cell-based detection | Use of cells for detecting unknown toxicants |
| | Cell/viral components | Viro-wires, protein capacitors/batteries, viral batteries |
| | Acoustic ranging | Ultrasonic echolocation based on bats |
| System | Biocomputing | Development of DNA computing, genetic algorithms, memory etc. |
| | Bioenergy systems | Energy harvesting systems, biological photovoltaic systems |
| | Bio-robotics | Land mobile systems using biological principles |
| | Bio-micro-air vehicles | Biological-inspired flapping wings system, flight path control/sensing |
| | Bio-efficient aircraft | Exploiting how birds fly and navigate |
| | Bio-submersibles | Aquatic systems based on fish, octopi etc. |
| System of systems | Network-enabled capability | Connecting information from a wide range of differing sensing platforms |
| | Multiple autonomous environment-specific platforms | Collective/swarming/shoaling behaviours for a variety of autonomic platforms |
| | Health monitoring | Systems that provide status of the individual's health whether a civilian or soldier |
| | Cybersecurity | Bio-inspired approaches based on the immune system for security |

properties. An example of the direct application of a biological entity is the use of virus particles as building blocks for materials and sensing systems.[4] Viruses offer a source of nanoscale particles that are monodisperse, that is, all the particles are of the same size. This is very difficult or almost impossible to achieve in conventional processes for creating nanoparticles. The molecules and structures on the surface of the virus particles, either naturally present or as a result of bioengineering, offer an opportunity to attach particles to each other or add specific desired molecules in particular places. Together, the uniform particle size and the complex multifunctional surfaces give many possibilities for creating new materials with interesting optical and other properties. Assembling ordered arrangements of particles whose size is similar to the wavelength of light leads to what is known as 'photonic band gap' materials. By manipulating the surface chemistry of virus particles they can be made to self assemble into strong macroscopic films or bulk materials, or be sensitive to chemical or biological toxicants, or to heat, pressure and other physical stimuli. The use of treated particles as sensing surfaces could also lead to bi-materials with stealth characteristics, to non-illuminating paints and coatings, or biological markers that can distinguish friendly soldiers.

The intricate structures and materials used by biological organisms results in enormous scope at the component level for a SASS system. For example, insect wing structures not only have ultra light aerodynamic designs but also employ surface nano-structuring of the materials to give distinctive colouring without the use of pigments. The Morpho butterfly is an excellent example (Figure 10.3). The colour stems from diffractive processes that manipulate the electromagnetic fields in ways that are not dissimilar to the photonic band-gap materials previously mentioned.

Structural materials, from the strength and toughness of wood and bone to protective armour of shell and the weaponry of tooth, horn and claw

**Figure 10.3** The Morpho Butterfly (Reproduced from http://commons.wikimedia.org/wiki/File:Blue_morpho_butterfly_300x271.jpg under a Creative Commons Attribution-Share Alike 3.0 Unported licence.)

demonstrate the efficiency of biological systems. What is remarkable is that these materials are assembled at the molecular level from simple materials with unremarkable properties but with an intricate hierarchical structure on length scales from nanometres to millimeters. The properties of spider silk are well known, and they result from the complex hierarchical structure of silk proteins. Biological systems can show extremes of performance in (reversible) adhesion, thermal management and in creating surfaces with super-hydrophilic or hydrophobic properties.

### 10.3.2 System Level

Many of the sensors (and counter sensors) found in natural systems are very elegant and have a level of complexity that we are only just beginning to understand. The ultrasonic system used by bats for night navigation and detecting prey is an excellent example of such a system. Bats emit radiation in the ultrasonic frequency range which illuminates the surroundings and the reflected radiation is picked up by their ears and processed. This is similar to a radar system. The bat uses the time between emitting the radiation and receiving the reflected radiation to determine distance. The processing time must be extremely short as the bat is flying and maneuvering rapidly and needs to process the information contained in the reflected radiation to update its distance from objects. The bat brain can also discriminate between stationary objects and those that are moving and between the wing beats of insects, such as butterflies, moths and mosquitoes and the movement of grass and leaves on trees. Remarkably, bats that feed on fish are able to detect perturbations of the water surface caused by fish.

A number of research groups have worked to develop systems with the capability to carry a significant weight of materiel and sensing systems over difficult terrain. The assessment of the way a dog walks has led to the design of robotic systems that mimic the locomotion of canines. A good example of this is the 'BigDog' programme funded by the US DARPA and developed by Boston Dynamics and associates. As shown in Figure 10.4, it is a stable quadruped robot, built in 2005, that uses canine leg movement to cross difficult terrain. It is powered by a two stroke 15HP engine. The engine drives the hydraulic leg actuators, and it uses a laser gyroscope for balance. It has stereo vision and a light detection and ranging (LIDAR) system. BigDog is capable of moving at 4mph, can climb a 35 degree slope and walk across rubble, muddy paths, in snow and water and carry 154.2 kg.[5] Although this is a mechanical device, it locomotive capability is inspired from the way canines move across terrain. The term "Bio-robotics' can be applied to this type of robotic system as it represents how robots simulate or mimic how natural systems move.

An area of growing interest is the development of bio-inspired micro air vehicles (MAV's) using flapping wings. Many current MAV's have fixed or rotary wing designs, low manoeuvrability and lack a stealth capability in confined spaces. Research has shown that flapping wing systems can have the same hovering capability as the hummingbird.

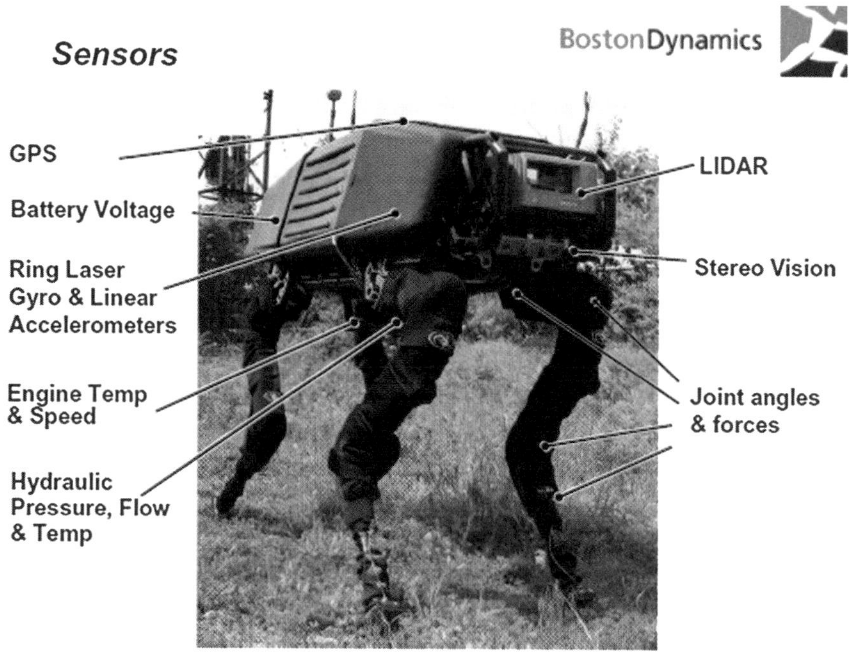
Sensors
BostonDynamics
GPS
Battery Voltage
Ring Laser Gyro & Linear Accelerometers
Engine Temp & Speed
Hydraulic Pressure, Flow & Temp
LIDAR
Stereo Vision
Joint angles & forces

**Figure 10.4** BigDog is a dynamically stable quadruped robot. (BigDog image courtesy of Boston Dynamics.)

As the field of bio-inspired robotic systems develops, one can envision systems that incorporate a range of bio-inspired sensors and seekers, cognitive capabilities, and the ability to harvest their energy from a variety of sources within the environment. This type of research is starting to show that conceptual models such as SASS can be turned into reality. It is also likely that as robotic systems develop they will also be able to scavenge for parts which will enable them to replicate. Initial research is underway on systems that can build copies of themselves (this level of research still requires the aid of human intervention). An example of this type of research involves RepRap[6] which is a free desktop 3-D printer capable of printing plastic objects. As many of the parts of RepRap are made from plastic, the RepRap machine can print its own parts but requires the human to assemble the 3-D printer.

## 10.3.3 System of Systems Level

The protection of strategically important buildings or facilities such as corporate offices, command and control centres is an example of the extension of the SASS model to the system of systems level. In this iteration of the concept, the main input and output services to a facility would be monitored for change or anomaly detection. The services monitored could include air ventilation, the mains water input and internal circulation. Other strategically important areas of the building such as doors, goods delivery points, command centres and conference rooms could also be monitored. Outer and inner rings of SASS nodes would form layers of protection around the most important regions of the facility. In such an application, the ability to identify the specific threat, such as chemical and biological agents or explosives would be secondary to identifying life-threatening changes in the environment. The analogy of the caged canary mentioned in Chapter 1, is apposite. The answer to the question, "What have we detected?' is not as important as the answer to the question, "What has changed from the normal known background environment that may be a threat to safety?"

If one considers sampling, collection and processing of a wide range of sensory inputs, humans in many situations are without parallel. When sailors doing rounds on a ship encounter a fire or a flood, they sample and collect a range of sensory inputs that lead them to the conclusion that there is or is not a problem in that space. These include visual, olfactory, thermal and auditory inputs. If a fire is present, smoke leads to obscuration of the space; flaming combustion presents another set of visual clues and olfaction indicates that thermal degradation products are present. The combustion process releases heat that can be sensed; and fires have an auditory component that varies depending on their size and the fuel source. A flood or pipe rupture in a space also presents a number of sensory clues that when processed by the brain allows an estimation of the severity of the situation. What is even more impressive is the ability that we have to differentiate between nuisance sources, such as burning toast or welding fumes and a real fire in a space or water being spilled from a bucket and flooding of a space due to a pipe rupture.

As navies move towards reduce crewing levels and ships are designed with more unmanned spaces, there is a need for a sensor system that can acquire input, process it and make decisions as to whether or not a situation in a space requires action to extinguish a fire or prevent flooding. The Naval Research Laboratory (NRL), Washington, DC has developed such a system.[7] The NRL volume sensor system consists of a sensor suite and data fusion and data analysis software to process the sensor input and determine if it is the result of an actual event (fire or flooding in a space) or if it is the result of other activities or events. The sensor suite includes a near infrared camera, a visible camera with image analysis capability, a microphone, and spectral sensors that operate in the ultraviolet and mid infrared range.

The key to success of such a system will be the speed of detection and the ability of the SASS nodes to communicate their status to either a central server or to each other for autonomous decision making. If the response time of the system is too slow, the benefits of early detection will be lost as the impact of the change will have already occurred. Hardwired communications between nodes which requires extensive cabling and maintenance could be impractical and expensive. Wireless communications make low powered node to node communication a more feasible option. This allows a configurable system capability, in which nodes can be moved to suit the changing environment. In addition, the ability of a node to provide an alarm gives directional information on the location and spread of a threat, allowing staff to move to safe areas or identify casualties. A network of SASS nodes means that failure of one node will not compromise the whole system, as communications between nodes can be easily re-routed. For example, if a node runs out of power or fails, the network can re-organize its intra-network communications to remain active. This collective behaviour approach provides redundancy and results in a more robust SASS network. By integrating the information provided by the sensors on each of the SASS nodes, minimal false alarm rates can be achieved. This enables low cost sensors to be used to provide an effective large area sensing network.

### 10.3.4 Range of Applications

A list of potential applications for the SASS concept is given in tables 3.1 & 10.1, the range of applications is considerable. The SASS concept could be used to overtly or covertly sense or monitor a variety of analytes or targets. Depending on its use, it could have no need for a power sources but be interrogated from a distance and its functional components could be biodegradable. The system could be sedentary, with a suite of sampling, transduction and processing elements, and have minimal power requirements that could be met by harvesting energy from the environment. It could be designed to be either sacrificial or retrievable. A more advanced system could be highly mobile and, with a suitable power source, carry a programmable range of sensing systems.

A SASS system could be designed to have networked enhanced capability, search for difficult targets or wounded soldiers, support the development of robust command and control systems, be autonomous and be used in search and rescue missions. However, to make best use of the growing wealth of bio-inspired and biotechnology research, there is a clear need to define both defence and security requirements, and to base them on technology that exists or is emerging.

## 10.4 Societal Implications

As bio-inspired interdisciplinary science starts to be exploited, there will be concerns about where this research and associated technological developments are leading. Significant developments in this area are happening around the globe. This is largely due to our growing understanding of cellular processes in living systems. With an increase in our ability to manipulate living systems at the molecular level and the growth in nanotechnology, the creation of bio-engineered systems could have tremendous benefit to societies. However, whilst these endeavours are for peaceful purposes, there is a potential for misuse. Hence, there will be a requirement for regulatory, ethical and legal policies to ensure transparency of research in these areas.

In May 2000, Matthew Meselson[8] warned about the potential dangers of biotechnology:

'Every major development - metallurgy, explosives, internal combustion, aviation, electronics, nuclear energy - has been intensively exploited, not only for peaceful purposes but also for hostile ones. Must this also happen with biotechnology, certain to be a dominant technology of the coming century? During the century just begun, as our ability to modify fundamental life processes continues its rapid advance, we will be able not only to devise additional ways to destroy life but . . . also . . . to manipulate it - including the processes of cognition, development, reproduction, and inheritance. A world in which these capabilities are widely employed for hostile purposes would be a world in which the very Nature of conflict has radically changed. Therein could lie unprecedented opportunities for violence, coercion, repression, or subjugation.'

Whilst advances in biotechnology offer to enhance living standards and national security, the dual use Nature of this technology cannot be ignored. As the fields of biotechnology and bio-inspiration expand and develop, there is a strong need to develop appropriate codes of conduct and bioethical frameworks to ensure that the benefits derived from this work are not put to misuse. There is a need for an overarching bioethical agreement between nations concerning this area.

Biotechnology and bio-inspired design can be used for materials and sensing systems with the potential to bring considerable benefit to all. It is up to the individuals involved to ensure transparency of the research and peacefully development of the resultant technologies.

## 10.5 Concluding Remarks

It may be that we will never be able to completely realise the promise of using bio-inspired approaches to produce materials, systems and system of systems that truly mimic those found in Nature. However, the ever-expanding body of knowledge in the area of bio-inspired materials and sensing systems has been and will continue to be used to produce novel materials and sensors that have properties or characteristics superior to those that they are replacing. By gaining a basic understanding of how Nature produces materials and components, then integrates them into functional components and systems, it may be possible to engineer systems with enhanced capabilities (sensitivity, selectivity and environmental resistance), reduced power requirements and cost, and improved processing and communications, situational awareness and robustness. The SASS concept provides a framework that can be used to identify weaknesses in existing systems and inform scientists and engineers where bio-inspired advances in the area of materials, collection, sensing, transduction, processing, communication and power can impact future sensing systems. It also highlights the requirements for an integrated approach to the fabrication of both components and complete sensing systems.

For the future, wide acceptance of bio-inspired approaches and the insertion of technologies arising from these discoveries into operational environments will face considerable challenges. To meet these challenges there is a clear need for interdisciplinary, cross-cutting interactions to integrate science and technology, design, fabrication and manufacturing with user requirements. Defence and security communities will need to be part of this process to avoid technological surprise.

# References


1. www.sciencemag.org/cgi/content/abstract/1173759
2. www.jcvi.org/cms/press/press-releases/full-text/article/first-self-replicating-synthetic-bacterial-cell-constructed-by-j-craig-venter-institute-research/ accessed 26th June 2010
3. www.aber.ac.uk/en/cs/research/cb/projects/robotscientist
4. X. D. Yan, N. H. Olson, J. L. Van Etten, M. Bergoin, M. G. Rossman and T. S. Baker, *Nature Struct. Biol.*, 2000, **7**, 101.
5. www.bostondynamics.com/img/BigDog_IFAC_Apr-8-2008.pdf
6. http://reprap.org/wiki/Main_Page
7. J. A. Lynch, D. T. Gottuk, J. C. Owrutsky, D. A. Minor, S. C. Wales, J. P. Farley, S. L. Rose-Pehrsson and F. W. Williams, Volume Sensor Development Test Series 5- Multi-Component System, NRL/MR/6180-05-8931, December 30 2005.
8. US National Academies of Science committee report on Biotechnology Research in an Age of Terrorism (ISBN 0-309-52613-2), www.nap.edu/catalog/10827.html

# Subject Index